MATHEMATICS
Coaching
&
Collaboration
in a PLC at Work®

Timothy D. Kanold   **Mona Toncheff**

Matthew R. Larson   Bill Barnes

Jessica Kanold-McIntyre   Sarah Schuhl

Solution Tree | Press

a division of
Solution Tree

555 North Morton Street
Bloomington, IN 47404
800.733.6786 (toll free) / 812.336.7700
FAX: 812.336.7790

email: info@SolutionTree.com
SolutionTree.com

Visit **go.SolutionTree.com/MathematicsatWork** to download the free reproducibles in this book.

Printed in the United States of America

Library of Congress Cataloging-in-Publication Data

Names: Kanold, Timothy D., author.
Title: Mathematics coaching and collaboration in a PLC at work / Timothy D.
  Kanold [and five others].
Other titles: Mathematics coaching and collaboration in a professional
  learning community at work
Description: Bloomington, IN : Solution Tree Press, [2018] | Includes
  bibliographical references and index.
Identifiers: LCCN 2017046682 | ISBN 9781943874347 (perfect bound)
Subjects: LCSH: Mathematics--Study and teaching. | Mathematics
  teachers--Training of.
Classification: LCC QA11.2 .G7425 2018 | DDC 510.71/2--dc23 LC record available at
https://lccn.loc.gov/2017046682

Solution Tree
Jeffrey C. Jones, CEO
Edmund M. Ackerman, President

Solution Tree Press
President and Publisher: Douglas M. Rife
Editorial Director: Sarah Payne-Mills
Art Director: Rian Anderson
Managing Production Editor: Caroline Cascio
Senior Production Editor: Suzanne Kraszewski
Senior Editor: Amy Rubenstein
Copy Editor: Ashante K. Thomas
Proofreader: Elisabeth Abrams
Text and Cover Designer: Laura Cox
Editorial Assistants: Jessi Finn and Kendra Slayton

Acknowledgments

Timothy D. Kanold

First and foremost, my thanks to each author on our team for understanding the joy, pain, and hard work of the writing journey, and for giving freely of their deep mathematics talent to so many others. Sarah, Matt, Bill, Mona, and especially my daughter, Jessica, you are each a gifted colleague and friend.

My thanks also to our reviewers, colleagues who have dedicated their lives to the work and effort described within the pages of this assessment book, and especially Kit, Sharon, and Jenn, who agreed to dedicate the time necessary to review each book in the series. I personally am grateful to each reviewer, and I know that you also believe that *every student can learn mathematics.*

Thanks, too, to Jeff Jones, Douglas Rife, Suzanne Kraszewski, and the entire editorial team from Solution Tree for their belief in our vision and work in K–12 mathematics education and for making our writing and ideas so much better.

Thanks also to my wife, Susan, and the members of our "fambam" who understand how to love me and formatively guide me through the good and the tough times a series project as bold as this requires.

And finally, my thanks go to you, the reader. May you rediscover your love for mathematics and teaching each and every day. The story of your life work matters.

Mona Toncheff

During my educational career, I have had the privilege of teaching students with diverse learning needs and backgrounds. Each year, my goal is to make mathematics more accessible to inspire students to continue their mathematics journey.

Now, I have the privilege of working with teachers and leaders across the nation whose goal is the same. The teachers and leaders with whom I have learned over the past twenty-five years have inspired this book series. The vision for a deeper understanding of mathematics for each and every student is achievable when we collectively respond to diverse student learning.

I am forever grateful to my husband, Gordon, and team Toncheff, who always support and encourage my professional pursuits.

Finally, thank you Tim for your leadership, mentorship, and vision for this series; and to Sarah, Matthew, Bill, and Jess; I am a better leader because of learning and leading with you.

Matthew R. Larson

It was an honor to be a member of an authorship team that truly understands each and every student can learn mathematics at a high level if the necessary conditions are in place. I thank all of you for your leadership and commitment to students and their teachers.

My thanks also go out to the thousands of dedicated teachers of mathematics who continue their own learning and work tirelessly every day to reach their students to ensure they experience mathematics in ways that prepare them for college and careers, promote active engagement in our democratic society, and help them experience the joy and wonder of learning mathematics. I hope this book will support you in your critical work.

Bill Barnes

First and foremost, thank you to our authorship team, Sarah, Mona, Jessica, and Matt, colleagues and friends who helped me overcome challenges associated with writing about the deep-rooted culture associated with homework and grading practices. Special thanks to Tim Kanold for continuing to believe in me, for supporting my work, and for helping me find my voice.

Special thanks to my wife, Page, and my daughter, Abby, who encourage me to pursue a diverse set of professional and personal interests. I am so grateful for your love, support, and patience as I continue to grow as a husband and father.

Finally, thanks to my work family, friends, and colleagues from the Howard County Public School System. I would like to especially thank our superintendent, Dr. Michael J. Martirano, for insisting that I follow my passion and continue to grow as a professional.

Jessica Kanold-McIntyre

I am extremely grateful to the writing team of Bill, Mona, Sarah, Tim, and Matt for their inspiration and collaboration through this writing process. Thank you to the teachers and leaders that have encouraged and influenced me along my own professional journey. Your passion and dedication to students and student learning continue to help me grow and learn. A special thanks to my husband for his support and encouragement throughout the process of writing this book. And, last but not least, I'd like to acknowledge our two dogs, Mac and Lulu, and my daughter Abigail Rose, who was born during the writing of this book, for their constant love.

Sarah Schuhl

Working to advance mathematics education within the context of a professional learning community can be a daunting task. Many thanks to Tim for his vision, encouragement, feedback, and support. Since our paths first crossed, he has become my mentor, colleague, and friend. Your heart for students and educators is a beautiful part of all you do.

Additionally, thank you to my dear friends and colleagues Mona, Bill, Jessica, and Matt, led by Tim in his team approach to completing this series. I am grateful to work with such a talented team of educators who are focused on ensuring the mathematics learning of each and every student. You have challenged me and made the work both inspiring and fun.

To the many teachers I have worked with, thank you for your tireless dedication to students and your willingness to let me learn with you. A special thank you to the teachers at Aloha-Huber Park School for working with me to give tasks so we could gather evidence of student reasoning and thinking to share.

Finally, I was only able to be a part of this series because of the support, love, and laughter given to me by my husband, Jon, and our sons, Jacob and Sam. Your patience and encouragement mean the world to me and will be forever appreciated.

Solution Tree Press would like to thank the following reviewers:

Jennifer Deinhart
K–8 Mathematics Specialist
Mason Crest Elementary School
Annandale, Virginia

Ellen Delaney
Executive Director
Minnesota Council of Teachers of Mathematics
St. Paul, Minnesota

Darshan M. Jain
Director of Mathematics and Computer Science
Adlai E. Stevenson High School
Lincolnshire, Illinois

Kit Norris
Educational Consultant
Hudson, Massachusetts

Sharon Rendon
Mathematics Consultant
National Council of Supervisors of Mathematics
Central 2 Regional Director
Rapid City, South Dakota

Denise Walston
Director of Mathematics
Council of the Great City Schools
Washington, DC

Jon Yost
Assistant Superintendent of Collaboration and Leadership
Sanger Unified School District
Sanger, California

Visit **go.SolutionTree.com/MathematicsatWork** to download the free reproducibles in this book.

Table of Contents

About the Authors

Timothy D. Kanold, PhD, is an award-winning educator, author, and consultant and national thought leader in mathematics. He is former director of mathematics and science and served as superintendent of Adlai E. Stevenson High School District 125, a model professional learning community (PLC) district in Lincolnshire, Illinois.

Dr. Kanold is committed to equity and excellence for students, faculty, and school administrators. He conducts highly motivational professional development leadership seminars worldwide with a focus on turning school vision into realized action that creates greater equity for students through the effective delivery of the PLC process by faculty and administrators.

He is a past president of the National Council of Supervisors of Mathematics (NCSM) and coauthor of several best-selling mathematics textbooks over several decades. Dr. Kanold has authored or coauthored thirteen books on K–12 mathematics and school leadership since 2011, including the bestselling book *HEART!* He also has served on writing commissions for the National Council of Teachers of Mathematics (NCTM) and has authored numerous articles and chapters on school leadership and development for education publications since 2006.

Dr. Kanold received the 2017 Ross Taylor/Glenn Gilbert Mathematics Education Leadership Award from the National Council of Supervisors of Mathematics, the international 2010 Damen Award for outstanding contributions to the leadership field of education from Loyola University Chicago, 1986 Presidential Awards for Excellence in Mathematics and Science Teaching, and the 1994 Outstanding Administrator Award from the Illinois State Board of Education. He serves as an adjunct faculty member for the graduate school at Loyola University Chicago.

Dr. Kanold earned a bachelor's degree in education and a master's degree in mathematics from Illinois State University. He also completed a master's degree in educational administration at the University of Illinois and received a doctorate in educational leadership and counseling psychology from Loyola University Chicago.

To learn more about Timothy D. Kanold's work, visit his blog, *Turning Vision Into Action* (www.turningvisionintoaction.today) and follow him on Twitter @tkanold.

Mona Toncheff, an educational consultant and author, is also currently working as a project manager for the Arizona Mathematics Partnership (a National Science Foundation–funded grant). A passionate educator working with diverse populations in a Title I district, she previously worked as both a mathematics teacher and a mathematics content specialist for the Phoenix Union High School District in Arizona. In the latter role, she provided professional development to high school teachers and administrators related to quality mathematics teaching and learning and working in effective collaborative teams.

As a writer and consultant, Mona works with educators and leaders nationwide to build collaborative teams, empowering them with effective strategies for aligning curriculum, instruction, and assessment to ensure all students receive high-quality mathematics instruction.

Toncheff is currently an active member of the National Council of Supervisors of Mathematics (NCSM) board and has served NCSM in the roles of secretary (2007–2008), director of western region 1 (2012–2015), second vice president (2015–2016), first vice president (2016–2017), marketing and e-news editor (2017–2018), and president-elect (2018–2019). In addition to her work with NCSM, Mona is also the current president of Arizona Mathematics Leaders (2016–2018). She was named 2009 Phoenix Union High School District Teacher of the Year; and in 2014, she received the Copper Apple Award for leadership in mathematics from the Arizona Association of Teachers of Mathematics.

Toncheff earned a bachelor of science degree from Arizona State University and a master of education degree in educational leadership from Northern Arizona University.

To learn more about Mona Toncheff's work, follow her on Twitter @toncheff5.

Matthew R. Larson, PhD, is an award-winning educator and author who served as the K–12 mathematics curriculum specialist for Lincoln Public Schools in Nebraska for more than twenty years. He served as president of the National Council of Teachers of Mathematics (NCTM) from 2016–2018. Dr. Larson has taught mathematics at the elementary through college levels and has held an honorary appointment as a visiting associate professor of mathematics education at Teachers College, Columbia University.

He is coauthor of several mathematics textbooks, professional books, and articles on mathematics education, and was a contributing writer on the influential publications *Principles to Actions: Ensuring Mathematical Success for All* (NCTM, 2014) and *Catalyzing Change in High School Mathematics: Initiating Critical Conversations* (NCTM, 2018). A frequent keynote speaker at national meetings, Dr. Larson's humorous presentations are well-known for their application of research findings to practice.

Dr. Larson earned a bachelor's degree and doctorate from the University of Nebraska–Lincoln, where he is an adjunct professor in the department of mathematics.

Bill Barnes is the chief academic officer for the Howard County Public School System in Maryland. He is also the second vice president of the NCSM and has served as an adjunct professor for Johns Hopkins University, the University of Maryland–Baltimore County, McDaniel College, and Towson University.

Barnes is passionate about ensuring equity and access in mathematics for students, families, and staff. His experiences drive his advocacy efforts as he works to ensure opportunity and access to underserved and underperforming populations. He fosters partnership among schools, families, and community resources in an effort to eliminate traditional educational barriers.

A past president of the Maryland Council of Teachers of Mathematics, Barnes has served as the affiliate service committee eastern region 2 representative for the NCTM and regional team leader for the NCSM.

Barnes is the recipient of the 2003 Maryland Presidential Award for Excellence in Mathematics and Science Teaching. He was named Outstanding Middle School Math Teacher by the Maryland Council of Teachers of Mathematics and Maryland Public Television and Master Teacher of the Year by the National Teacher Training Institute.

Barnes earned a bachelor of science degree in mathematics from Towson University and a master of science degree in mathematics and science education from Johns Hopkins University.

To learn more about Bill Barnes's work, follow him on Twitter @BillJBarnes.

Jessica Kanold-McIntyre is an educational consultant and author committed to supporting teacher implementation of rigorous mathematics curriculum and assessment practices blended with research-informed instructional practices. She works with teachers and schools around the country to meet the needs of their students. Specifically, she specializes in building and supporting the collaborative teacher culture through the curriculum, assessment, and instruction cycle.

She has served as a middle school principal, assistant principal, and mathematics teacher and leader. As principal of Aptakisic Junior High School in Buffalo Grove, Illinois, she supported her teachers in implementing initiatives, such as the Illinois Learning Standards; Next Generation Science Standards; and the College, Career, and Civic Life Framework for Social Studies State Standards, while also supporting a one-to-one iPad environment for all students. She focused on teacher instruction through the PLC process, creating learning opportunities around formative assessment practices, data analysis, and student engagement. She previously served as assistant principal at Aptakisic, where she led and supported special education, response to intervention (RTI), and English learner staff through the PLC process.

As a mathematics teacher and leader, Kanold-McIntyre strived to create equitable and rigorous learning opportunities for all students while also providing them with cutting-edge 21st century experiences that engage and challenge them. As a mathematics leader, she developed and implemented a districtwide process for the Common Core State Standards in Illinois and led a collaborative process to create mathematics curriculum guides for K–8 mathematics, algebra 1, and algebra 2. She currently serves as a board member for the National Council of Supervisors of Mathematics (NCSM).

Kanold-McIntyre earned a bachelor's degree in elementary education from Wheaton College and a master's degree in educational administration from Northern Illinois University. To learn more about Jessica Kanold-McIntyre's work, follow her on Twitter @jkanold.

Sarah Schuhl is an educational coach and consultant specializing in mathematics, professional learning communities, common formative and summative assessments, school improvement, and response to intervention (RTI). She has worked in schools as a secondary mathematics teacher, high school instructional coach, and K–12 mathematics specialist.

Schuhl was instrumental in the creation of a PLC in the Centennial School District in Oregon, helping teachers make large gains in student achievement. She earned the Centennial School District Triple C Award in 2012.

Sarah designs meaningful professional development in districts throughout the United States focused on strengthening the teaching and learning of mathematics, having teachers learn from one another when working effectively as a collaborative team in a PLC, and striving to ensure the learning of each and every student through assessment practices and intervention. Her practical approach includes working with teachers and administrators to implement assessments for learning, analyze data, collectively respond to student learning, and map standards.

Since 2015, Schuhl has coauthored the books *Engage in the Mathematical Practices: Strategies to Build Numeracy and Literacy With K–5 Learners* and *School Improvement for All: A How-to Guide for Doing the Right Work.*

Previously, Schuhl served as a member and chair of the National Council of Teachers of Mathematics (NCTM) editorial panel for the journal *Mathematics Teacher.* Her work with the Oregon Department of Education includes designing mathematics assessment items, test specifications and blueprints, and rubrics for achievement-level descriptors. She has also contributed as a writer to a middle school mathematics series and an elementary mathematics intervention program.

Schuhl earned a bachelor of science in mathematics from Eastern Oregon University and a master of science in mathematics education from Portland State University.

To learn more about Sarah Schuhl's work, follow her on Twitter @SSchuhl.

To book Timothy D. Kanold, Mona Toncheff, Matthew R. Larson, Bill Barnes, Jessica Kanold-McIntyre, or Sarah Schuhl for professional development, contact pd@SolutionTree.com.

Preface

By Timothy D. Kanold

In the early 1990s, I had the honor of working with Rick DuFour at Adlai E. Stevenson High School in Lincolnshire, Illinois. During that time, Rick—then principal of Stevenson—began his revolutionary work as one of the architects of the Professional Learning Communities at Work® (PLC) process. My role at Stevenson was to initiate and incorporate the elements of the PLC process into the K–12 mathematics programs, including the K–5 and 6–8 schools feeding into the Stevenson district. Like you, I had entered into the difficult challenge of leading other teachers of mathematics in their daily work.

In those early days of our PLC pursuit, we understood the grade-level or course-based mathematics collaborative teacher team provided us a chance to share and become more transparent with one another. We exchanged knowledge about our own growth and improvement as teachers of mathematics and began to create and enhance student agency for learning mathematics. This meant we needed to understand ways in which our mathematics content, instruction, and assessment design facilitated student *ownership* in their learning, their meaning making, and their actions toward learning mathematics. As colleagues and team members, we became more open and taught, coached, argued, and learned from one another.

And yet, we did not know our mathematics *coaching collaboration* story. And for sure, we did not know how to lead the mathematics PLC process effectively. We did not have much clarity on how to reach and lead teacher team agreements on mathematics homework assignments, test questions, use of high- and low-cognitive-demand tasks during a lesson, and the grading and scoring of student work. These had all been private affairs of our previous professional actions.

We did not *yet* understand the idea that coaching and collaborating with one another would improve the learning of mathematics for *all* students. In our minds we were self-sufficient, and we left each other alone. We were mostly self-focused on our own students. The idea of driving to work and working *together* to care for *all* of the students in a grade level or math course and not just for the ones assigned to us presented a new vision for our professional work life.

We did not initially understand how our daily work as part of a collaborative teacher team in mathematics at all grade levels could erase inequities in student learning caused by the wide variance of our professional and isolated teaching practices. In most cases, we, not our students, were causing the gaps in learning mathematics. And no one was leading us forward to address these inequities together.

Ultimately, leading the effort to close the student and adult learning gaps became my job at Stevenson. This became the job of lead coauthor Mona Toncheff in Phoenix, Arizona; coauthor and researcher Matt Larson in Lincoln, Nebraska; coauthor Sarah Schuhl in Portland, Oregon; coauthor Jessica Kanold-McIntyre in suburban Chicago; and coauthor Bill Barnes in Howard County, Maryland. We have spent significant time in the classroom as highly successful practitioners, leaders, and coaches of K–12 mathematics teams, designing and leading the structures and the culture necessary for effective collaborative team efforts. We have lived through and led the mathematics professional growth actions this book advocates within diverse K–12 educational settings in rural, urban, and suburban schools. We have each been in your K–12 leadership shoes as math coaches and as school administrators. We are practitioners in the field. And that is why we know the collaboration tools and protocols we provide in this book will be an asset to your mathematics leadership work.

During my years at Stevenson, our faculty could not have anticipated one of the best benefits of working in community with one another: the benefit of belonging to something larger than ourselves. There is a benefit to learning about various teaching and assessing strategies from each other, *as professionals*. We realized it was often in community we found deeper meaning to our work, and strength in the journey as we solved the complex issues we faced each day and each week of the school season, *together*.

As we began our collaborative mathematics work at Stevenson and with our feeder districts, we discovered quite a bit about our mathematics teaching and learning leadership. And remember, we were doing this work together in the early 1990s—well before the idea of transparency in practice and observing and learning from one another in our professional work became as popular as they are today.

We discovered that if we were to become a professional *learning* community, then experimenting together and trying out effective strategies of practice needed to become our norm. We needed to learn more about the mathematics curriculum, the nature of the mathematics tasks we were choosing each day, and the types of lessons and assessments we were developing and using. And, we realized we needed to do so *together*.

In this K–12 *Every Student Can Learn Mathematics* series, we emphasize the concept of *team action*. We recognize that some readers may be the only members of a grade level or mathematics course. In that case, we recommend you help them work with a colleague at a grade level or course above or below their own. Or, help them to work with other job-alike teachers across a geographical region as technology allows. Your mathematics collaborative teams are the engines that will drive your mathematics program success.

A PLC in its truest form is "an ongoing process in which educators work collaboratively in recurring cycles of collective inquiry and action research to achieve better results for the students they serve" (DuFour, DuFour, Eaker, Many, & Mattos, 2016, p. 10). This book and the other three in the K–12 *Every Student Can Learn Mathematics* series feature a wide range of voices, tools, and discussion protocols offering advice, tips, and knowledge for your PLC-based collaborative mathematics teams.

In this book, we tell our mathematics coaching and collaboration story. It is a K–12 story that when well implemented, will bring great satisfaction to your work as a mathematics professional and leader and result in a positive impact on your teachers and your students.

We hope you will join us in the journey of significantly improving teacher and student learning in mathematics by leading and improving the collaboration story for each of your teams, your schools, or your districts you lead. The conditions and the actions for adult learning of mathematics *together* are included in these pages. We hope the stories we tell, the tools we provide, and the opportunities for reflection serve you well in your daily leadership work in a discipline we all love—mathematics!

Introduction

If you are a teacher leader of mathematics, then this book is for you! Whether you are a novice or a master teacher; an elementary, middle, or high school teacher leader; a rural, suburban, or urban teacher leader; this book is for you. It is for all school site administrators, teacher team leaders, instructional coaches, and central office leaders who support professionals who are part of the K–12 mathematics learning experience.

Leading mathematics so *each and every student* learns the K–12 college-preparatory mathematics curriculum, develops a positive mathematics identity, and becomes empowered by mathematics is a complex and challenging task. Trying to tackle that task by allowing your teachers to work in isolation from their colleagues will severely limit your ability to erase inequities in student learning that exist in your schools. The pursuit and hope of developing your teachers into a collaborative community with their colleagues and moving them away from isolated professional practice are necessary, hard, exhausting, and sometimes overwhelming.

In this mathematics leadership book, you and your colleagues will learn to focus your time and energy on teaching others how their collaborative efforts will result in significant improvement in student learning and promote equity.

What is equity? To answer that, it is helpful to first examine inequity. In traditional schools in which teachers work in isolation, there is often a wide discrepancy in teacher practice. Teachers in the same grade level or course may teach and assess mathematics quite differently—there may be a lack of consistency in what your grade-level or course-based teachers expect students to know and be able to do, how they will know when students have learned, what they will do when students have not learned, and how they will proceed when students have demonstrated learning. Such wide variance in potential teacher practice among grade-level and course-based teachers then causes inequities in the student learning of mathematics as students pass from course to course and grade to grade. The gap widens.

Equity and PLCs

The good news is you can lead the erasing of the mathematics learning gaps. The PLC at Work® process is one of the best and most promising models your school or district can use to build a more equitable response for student learning. The architects of the PLC process, Richard DuFour, Robert Eaker, and Rebecca DuFour, designed the process around three big ideas and four critical questions that placed learning, collaboration, and results at the forefront of our work (DuFour et al., 2016). School and district leadership that commit to the PLC transformation process rally around the following three big ideas (DuFour et al., 2016).

1. **A focus on learning:** Teachers focus on learning as the fundamental purpose of the school rather than on teaching as the fundamental purpose.

2. **A collaborative culture:** Teachers work together in teams interdependently to achieve a common goal or goals for which members are mutually accountable.

3. **A results orientation:** Team members are constantly seeking evidence of the results they desire—high levels of student learning.

Additionally, you lead the mathematics teacher teams within a PLC focus on four critical questions (DuFour et al., 2016):

1. What do we want all students to know and be able to do?

2. How will we know if they learn it?

3. How will we respond when some students do not learn?

4. How will we extend the learning for students who are already proficient?

The four critical PLC questions provide an equity lens for your professional work. Imagine the opportunity gaps that will exist if you do not help your colleagues to agree on the level of rigor for PLC critical question 1, What do we want all students to know and be able to do?

Imagine the devastating effects on students if you do not help your faculty reach complete team agreement on the high-quality criteria for the assessments you administer (PLC critical question 2) and the routines your teachers use for how you score those mathematics assessments. Imagine the lack of student agency (their voice in learning) if you do not help the teachers learn how to work together to create a unified, robust formative mathematics assessment process for helping students *own* their response when they are and are not learning.

To answer these four PLC critical questions well requires structures through the development of the *products* you help teams to produce together, and a formative culture through the *process* of how you work with your teams to *use* those products.

The concept of your mathematics teacher teams *reflecting together and then taking action* around the right work is an emphasis in the K–12 *Every Student Can Learn Mathematics* series. The potential actions you and your colleagues take together improve the likelihood of more equitable mathematics learning experiences for every K–12 student. And you must help to lead the teacher teams forward around the right professional work.

The Reflect, Refine, and Act Cycle

Figure I.1 illustrates our author perspective about the process of lifelong learning—for us, for you, and for your students. The very nature of our profession—education—is about the development of skills toward learning. Those skills are part of an ongoing process we pursue together.

More important, the reflect, refine, and act cycle is a *formative* learning cycle we describe throughout all four books in the series. When you embrace mathematics learning as a *process*, you and your students:

- **Reflect**—Work the task, and then ask, "*Is this the best solution strategy?*"

- **Refine**—Receive FAST Feedback and ask, "*Do I embrace my errors?*"

- **Act**—Persevere and ask, "*Do I seek to understand my own learning?*"

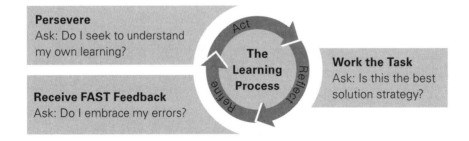

Figure I.1: Reflect, refine, and act cycle for formative student learning.

The intent of the *Every Student Can Learn Mathematics* series is to provide you with a systemic way to structure and facilitate the deep team discussions necessary to lead an effective and ongoing adult and student learning process each and every school year.

Mathematics in a PLC at Work Framework

The *Every Student Can Learn Mathematics* series has four books that focus on a total of six teacher team actions and two mathematics coaching actions within four primary categories.

1. *Mathematics Assessment and Intervention in a PLC at Work*

2. *Mathematics Instruction and Tasks in a PLC at Work*

3. *Mathematics Homework and Grading in a PLC at Work*

4. *Mathematics Coaching and Collaboration in a PLC at Work*

Figure I.2 (page 4) shows each of these four categories and the two actions within them. These eight actions focus on the nature of the professional work of your teacher teams and how they should respond to the four critical questions of a PLC (DuFour et al., 2016).

This book in the series provides a leadership focus on your coaching actions so that the six teacher team actions can become a reality. Your work begins by deciding who exactly should be working in your collaborative teams for mathematics to develop the team actions described in figure I.1.

Most commonly, your collaborative teams will consist of two or more teachers who teach the same grade level or course, and the support personnel such as resource teachers for those students. Through your focused work in helping your teachers address the four critical questions of a PLC, you provide every student in your grade-level or course-based teams with equitable learning experiences and expectations, opportunities for sustained perseverance, and robust formative feedback, regardless of the teacher he or she receives.

If, however, there are singleton (lone teachers who do not have a colleague who teaches the same grade level or course) teachers in your school, you will have to determine who it makes the most sense for them to

work with as you strengthen their lesson-design and student-feedback skills. Leadership consultant and author Aaron Hansen (2015) suggests the following possibilities for creating teams for singletons.

- Vertical teams (for example, a primary school team of grades K–2 teachers or a middle school mathematics department team for grades 6–8)

- Virtual teams (for example, a team comprising teachers from different sites who teach the same grade level or course and collaborate virtually with one another across geographical regions)

- Grade-level or course-based teams (for example, a team of grade-level or course-based teachers in which each teacher teaches all sections of grade 6, grade 7, or grade 8; the teachers might expand to teach and share two or three grade levels instead of only one to create grade-level or course-based teams)

About This Book

In *Mathematics Coaching and Collaboration in a PLC at Work*, we explore two specific actions for your professional work as a coach within the collaborative culture of a PLC.

1. **Coaching action 1:** Develop PLC structures for effective teacher team engagement, transparency, and action.

2. **Coaching action 2:** Use common assessments and lesson-design elements for teacher team reflection, data analysis, and subsequent action.

In the chapters of part 1 of this book, you will explore different leadership development practices and several leadership strategies or actions. You will discover how becoming strong in these practices and strategies promotes effective team leadership. In part 2, you will examine numerous protocols to support teacher team reflection and continual growth by helping your mathematics teacher teams to engage in *action research*—a cycle of analyzing evidence of student learning during and after a unit of mathematics instruction and assessment.

So how do you coach and lead meaningful change and improvement initiatives in mathematics at the classroom level? This book provides a guide to coaching actions that provide the greatest benefit for sustaining deep change with each mathematics team and includes

Every Student Can Learn Mathematics Series Team and Coaching Actions Serving the Four Critical Questions of a PLC at Work	1. What do we want all students to know and be able to do?	2. How will we know if students learn it?	3. How will we respond when some students do not learn?	4. How will we extend the learning for students who are already proficient?
Mathematics Assessment and Intervention in a PLC at Work				
Team action 1: Develop high-quality common assessments for the agreed-on essential learning standards.	■	■		
Team action 2: Use common assessments for formative student learning and intervention.			■	■
Mathematics Instruction and Tasks in a PLC at Work				
Team action 3: Develop high-quality mathematics lessons for daily instruction.	■	■		
Team action 4: Use effective lesson designs to provide formative feedback and student perseverance.			■	■
Mathematics Homework and Grading in a PLC at Work				
Team action 5: Develop and use high-quality common independent practice assignments for formative student learning.	■	■		
Team action 6: Develop and use high-quality common grading components and formative grading routines.			■	■
Mathematics Coaching and Collaboration in a PLC at Work				
Coaching action 1: Develop PLC structures for effective teacher team engagement, transparency, and action.	■	■		
Coaching action 2: Use common assessments and lesson-design elements for teacher team reflection, data analysis, and subsequent action.			■	■

Figure I.2: Mathematics in a PLC at Work framework.

Visit go.SolutionTree.com/MathematicsatWork for a free reproducible version of this figure.

insights and tools for lesson study, embedded coaching, and a response to evidence of student learning from the common unit assessments.

In this book, you will find discussion tools that offer questions for reflection, discussion, and action about your mathematics practices and routines. Visit **go.Solution Tree.com/MathematicsatWork** to find the free reproducibles listed in this book. In addition, you will find the Mathematics in a PLC at Work lesson-design tool used throughout this series. More important, you will find a comprehensive set of free online resources as a reference guide to support your work in leading mathematics teaching and learning every day, as well as a reference for print resources for your work as a mathematics education leader. You are invited to write your personal leadership story in the *leader reflection* boxes as well. To help you reflect on current progress in your collaborative team work that this book and the series as a whole describe, you can use the Mathematics in a PLC at Work Rating and Reflection Tool (page 109–110) to evaluate areas of strength and weakness in your current mathematics professional work and future efforts. You will also find some personal stories from the talented authors of this series. These stories provide a glimpse into the authors' personal insights, experiences, and practical advice connected to some of the strategies and ideas in this book.

The K–12 *Every Student Can Learn Mathematics* series is steeped in the belief that the decisions and daily actions of classroom teachers of mathematics matter.

Classroom teachers have the power to decide and choose the mathematical tasks students are required to complete during lessons, homework they develop and design, unit assessments such as quizzes and tests, and projects and other high-performance tasks. Your teachers have the power to determine the rigor for those mathematical tasks, the nature of student communication and discourse to learn those tasks, and whether or not learning mathematics should be a formative process for teachers and students. You, as a leader, must ensure the choices of the teachers are good choices, each and every day.

Most important, you have the power to decide if you and the classroom teachers in your school will do all of the challenging mathematics work of your profession alone or with others. As you embrace the belief that together, the work of your professional learning community can overcome the many obstacles you face each day, then the belief that *every student can learn mathematics* just may become a reality in your school.

Coaching Action 1:
Develop PLC Structures for Effective Teacher Team Engagement, Transparency, and Action

> The key to successful leadership is influence, not authority.
>
> —*Ken Blanchard*

You are most likely using *Mathematics Coaching and Collaboration in a PLC at Work* in the *Every Student Can Learn Mathematics* series because you are working hard to create a magnified impact for change in your mathematics program. You may be trying to erase the student inequities. Consider the following thoughts on equity from University of Illinois mathematics professor Rochelle Gutiérrez (2002):

> Equity in mathematics education will not be achieved until it is no longer possible to predict mathematics achievement and participation based solely on student characteristics such as race, class, ethnicity, sex, beliefs, and proficiency in the dominant language. (p. 153)

The road to creating a culture of equity is paved with the structures—the six team actions—we describe in the *Every Student Can Learn Mathematics* series (*Mathematics Assessment and Intervention in a PLC at Work*, *Mathematics Homework and Grading in a PLC at Work*, and *Mathematics Instruction and Tasks in a PLC at Work* [see Kanold, Barnes, et al., 2018; Kanold, Kanold-McIntyre, et al., 2018; and Kanold, Schuhl, et al., 2018]). When teams work *collaboratively* to make sense of the standards and make agreements on the assessments and interventions to provide each and every learner, they pave the way for erasing inequities in the student learning of mathematics.

This book serves then as a tool to provide you with a framework for leading, supporting, and monitoring the mathematics work of your collaborative teams as part of the PLC process in your school or district. As you lead your collaborative teacher teams, consider the following essential questions.

- Do my leadership practices engage others and inspire teacher transparency and collective action?
- Do my leadership strategies provide the necessary structures for effective teacher team collaboration?

Being a leader is about the engagement and inspiration of others. As school leaders, you have a sphere of influence that includes colleagues and supervisors to your north, south, east, and west (Kanold, 2011). Within this book, you explore the leadership practices you can cultivate in order to engage and inspire your collaborative team or teams within your sphere of influence.

To build a team where members are willing to learn from each other and engage in deeper conversations, your influence requires building trust and being mindful of the relational leadership strategies you use to establish a strong mathematics community. Your collaborative teams will trust the processes you employ when you communicate a clearly developed plan, as each teacher team member learns expectations for future actions and effort.

Building commitment for increased student understanding of mathematics depends on your ability to lead and coach others. Your leadership is vital for building collective efficacy and sets the tone and pace for how each team member engages in the collaborative process.

The work of mathematics grade-level or course-based collaborative teams is the driving force to ensuring each and every learner receives an equal opportunity to grow mathematically. Each collaborative team you lead demonstrates clarity about what students must learn, reaches agreement about how to assess that understanding, and collectively responds with its teaching peers when students do or do not learn.

When pursuing the PLC process, it becomes your expectation that each collaborative team will work interdependently to erase inequities in its practice and provide equitable learning experiences to ensure each and every student learns the mathematics standards for his or her grade level or course.

In the following sections, you will examine personal leadership development practices that will help you to personally grow as a leader of mathematics. You will also dig into leadership strategies you can use with your teacher teams.

You begin, however, with a discussion of mathematics leadership roles and how they need to work together to create an effective and coherent mathematics teacher team response in each school.

Mathematics Leadership Roles

As the National Council of Supervisors of Mathematics (NCSM, 2008) states, "A single mathematics education leader can have an incredible impact on the development and effectiveness of others" (p. 1). The collective leadership of your school or district impacts your individual collaborative teams and how effective each team will become. As a mathematics leader, you play a vital role in building the collective knowledge capacity of the teachers you serve. However, changing adults' behavior is not an easy task. Chip Heath and Dan Heath (2010) explain that for individuals' "behavior to change, you've got to influence not only their environment, but their hearts and mind" (p. 5). Simply identifying team leaders for collaborative teams, choosing a curriculum, or hiring a math coach for your site is not enough; each adult serving in a mathematics

leadership role should inspire one another as well. Take a few moments to reflect on the current mathematics leadership roles at your school site or district.

LEADER *Reflection*

What are the current mathematics leadership roles in your school or district? Who is providing leadership for the mathematics teaching and learning in your school?

_____

_____

_____

_____

_____

_____

_____

_____

_____

_____

_____

_____

_____

_____

_____

Leadership roles within schools and districts often fall into three categories.

1. Team leader

2. Mathematics coach

3. Site-level leader

Table P1.1 describes each of these three types. Which leadership role aligns with your current professional responsibilities?

Your mathematics leadership, regardless of your role, is vital as you are the builders of the collective knowledge capacity of the teachers you serve. The collective leadership and connections between these roles within your school or district impact your collaborative teams' performance and effectiveness.

Table P1.1: Mathematics Team Leadership Roles

Team Leader	Mathematics Coach	Site-Level Leader
• You orchestrate and facilitate the work of the collaborative team for mathematics. • You assist your grade-level or course-based mathematics team with making decisions and reaching agreements. • You are the point person to request support from coach or site-level leaders when challenges or questions arise with the mathematics program.	• You assist the mathematics team leader or team to navigate what is being done well and what needs to be done next. • You support the mathematics team leader and team members with connecting the work of their team (action steps and instructional focus) using individual coaching. • You provide job-embedded professional learning for mathematics.	• You provide feedback on the mathematics products and team processes teams use each week. • You grow in your own learning with the mathematics team leaders and coaches. • You provide nonevaluative support as your mathematics teams are learning. • You are the ultimate accountability person to ensure teams focus on the right work for student learning.

Leadership Practices and Strategies

The framework for leading effective mathematics collaborative teams includes developing your personal leadership practices by utilizing various leadership strategies with your teacher teams. *Practices* are different than *strategies*. The strategies provide the tools you need to engage in effective leadership practice development.

A *leadership practice* is a set of personal actions you intentionally develop through continuous reflection and training to improve your ability to lead (Kanold, 2011). A practice is a discipline with a variety of solution pathways you can use to support collaborative teams addressing the four critical questions of a PLC (Erkens & Twadell, 2012).

A leadership *strategy* is a protocol, tool, or process to make sense of something—such as the essential learning standards or during the creation of a common assessment (Erkens & Twadell, 2012).

Part 1 also includes leader and team discussion and reflection tools designed to support your work in Part 1.

- Coherence in Your Mathematics Leadership Practices (figure 1.1, page 13)

- Leadership Practices Rubric (figure 1.2, page 14)

- Sample Team Questions and Prompts (figure 1.5, page 19)

- Exploring Core Values Teacher Team Tool (figure 1.6, page 22)

- Productive and Unproductive Beliefs About Teaching and Learning Mathematics (figure 1.7, page 23)

- Leadership Strategies (figure 2.1, page 26)

- Leadership Strategies Scoring Rubric (figure 2.2, page 27)

- Grade 3 Mathematics Team Scenario for Common Purpose (figure 2.4. page 30)

- Guiding Questions to Develop Norms (figure 2.8, page 35)

- Survey of Team Norms (figure 2.9, page 35)

- SMART Goal and Data-Analysis-Planning Protocol (figure 2.10, page 37)

- Leading Team Reflections on Goal-Setting Actions (figure 2.11, page 39)

- Tight-Loose Aspects of Team Actions (figure 2.14, page 42)

- Mathematics Collaborative Team Reflection on the Three Big Ideas of a PLC Culture (figure 2.22, page 55)

Throughout this book, you will examine specific actions—intentional practices and strategies—to examine your collaborative teams' professional work. These practices and strategies address what all team members need to know and be able to do as part of the collaborative process in mathematics.

Table P1.2: Connections Between the Leadership Practices and Leadership Strategies

Five Leadership Practices Disciplines you develop and practice to support effective collaboration	Five Leadership Strategies Strategies you use to build specific leadership disciplines
1. **Trusting environment:** Your teams are willing to take risks and fully engage in the collaborative team process through a culture that honors trust.	1. **Common purpose:** You develop and articulate a shared mission and vision.
2. **Relational intelligence:** You lead and inspire by paying attention to the human aspect of your professional work. You use your emotional and relational intelligence to understand, learn from, and comprehend the interpersonal dynamics of your collaborative teams.	2. **Collective commitments:** You develop and articulate shared norms and values for how teacher team members will interact.
3. **Effective communication:** You use positive communication skills to empower the team and to build trust within the collaborative process. You listen without judgment to each and every team member. You value the person, even if you must help him or her change behaviors.	3. **Common goals:** You develop and articulate shared SMART (strategic and specific, measurable, attainable, results oriented, and time bound [Conzemius & O'Neill, 2014]) goals and action steps to drive the measurable action and work of the team.
4. **Passion and persistence:** You persevere through challenges by using your strong passion for mathematics leadership work.	4. **Protocols for continual reflection:** You use strategies that engage all team members in a reflect, refine, and act cycle of continuous improvement.
5. **Commitment to the PLC process:** The core values of the four critical questions of a PLC drive the work of your collaborative teams' commitments and beliefs.	5. **Mutual accountability:** You can collectively share the responsibilities of and accountability for improving the mathematics teaching and learning in your school or district.

Visit **go.SolutionTree.com/MathematicsatWork** *for a free reproducible version of this table.*

Table P1.2 represents the framework for the leadership practices and strategies you will explore in chapters 1 and 2.

In chapter 1, you will focus your professional work on the five leadership structures and personal practices *you* pursue in order to help your teacher teams be successful by helping each and every student to learn. In chapter 2, you will examine five leadership strategies for *how* to provide supportive conditions for effective collaboration. These strategies will support leading mathematics teams in collectively responding to student learning by coming to agreement on the standards all students are expected to learn and specific *measures of success* to gather and analyze throughout a unit of study.

CHAPTER 1

Five Personal PLC Leadership Development Practices

The true measure of leadership is influence, nothing more, nothing less.

—*John C. Maxwell*

Change is complex, and leaders like you, whose vision focuses on increased student mathematical proficiency, understand *what* your collaborative teams need to create as well as *how* your various grade-level or course-based mathematics teams engage in constant inquiry and action research. As noted in the introduction, the PLC is "an ongoing process in which educators work collaboratively in recurring cycles of collective inquiry and action research to achieve better results for the students they serve" (DuFour et al., 2016, p. 10).

The six collaborative team actions featured in the Mathematics in a PLC at Work Framework (figure I.2, page 4) and highlighted in the other three books in the *Every Student Can Learn Mathematics* series align to the three big ideas of a PLC culture and exemplify the attributes of a PLC (DuFour et al., 2016; Mattos, DuFour, DuFour, Eaker, & Many, 2016).

1. A focus on learning.

2. A collaborative culture.

3. A results orientation.

As a leader of mathematics in your school or district, you support the team process by creating a safe space for teacher team members to take risks, engage in productive struggle, make mistakes, and learn from their mistakes.

Not all mathematics teams are equal; some are definitely more productive and effective than others. Patrick Lencioni (2002) analyzes characteristics of dysfunctional teams and identifies a hierarchy of five dysfunctions. When teams have one dysfunction, that one dysfunction leads to another, and, ultimately, results in negative morale and ineffective teams.

LEADER *Reflection*

Think about a grade-level or course-based mathematics team you have recently been leading. What are the team's strengths?

What are some specific challenges the team faces? Are there specific adult behaviors that make the teamwork effective or other behaviors that cause dissonance on the team?

_____

_____

_____

_____

_____

_____

_____

Lencioni's (2002) dysfunctions include the following five.

1. **Absence of trust:** Trust is at the bottom of the hierarchy, and, when it is missing, team members are reluctant to take risks.

2. **Fear of conflict:** If there is a lack of trust, the fear of conflict emerges and team members are not comfortable stating their opinions to avoid any conflict.

3. **Lack of commitment**: If there is a fear of conflict, team members do not feel that their voice or opinions matter, and this will lead to a lack of commitment to the work of the team.

4. **Avoidance of accountability:** Due to the lack of commitment, team members do not make each other accountable for the team performance.

5. **Inattention to results:** If team members do not hold each other accountable, ultimately there is no focus on results or focus on learning.

When you develop the five leadership practices as described in table P1.2 (page 10), your purposeful actions support effective collaboration to overcome these typical dysfunctions. Fortunately, once you understand actions that do not support effective collaboration, you can engage in specific leadership practices and employ strategies to create more effective teams.

As you explore development in your personal leadership practices, take some time to reflect on the questions in figure 1.1. It will help you identify your current strengths and the challenges you face in your current leadership role, whether you are a team site-level leader or a mathematics instructional and assessment coach.

If your school site has an administrator, a mathematics coach and various mathematics team leaders and support personnel for mathematics, discuss your responses to figure 1.1 together.

You should then examine figure 1.2 (page 14) together, and use the criteria in the rubric to show that, in addition to understanding the work of collaborative teams, you also have to connect to team members' hearts and emotions to avoid the common dysfunctions Lencioni (2002) identifies. This requires you to understand the human side of effective collaboration—the emotions, passions, and energy required to serve others.

As described in figure 1.2, the five specific leadership practices you can develop and engage in as a leader to better support your collaborative teams are (1) trusting environment, (2) relational intelligence, (3) effective communication, (4) passion and persistence, and (5) commitment to the PLC process.

You can use figure 1.2 to rate your team or the teams you lead in each area. What is the total score on the rubric for you: fifteen, eight, twenty, or something else? In which of the criteria are your mathematics teams strongest? Which areas present challenges?

Team Leader	Mathematics Coach	Site-Level Leader
• What is the trust level on your mathematics team (low, medium, or high), and how do you build trust?	• Which mathematics teams have high levels of trust? Why? What do you believe are the contributing factors?	• Which mathematics teams have high levels of trust? Why? What do you believe are the contributing factors?
• How do you develop mutual respect among team members?	• Which of your teams support each other with mutual respect? What does that look like to you?	• Which of your teams support each other with mutual respect? What does that look like to you?
• How do you and your team consistently respond to the four critical questions of a PLC?	• What collegial actions do you observe that contribute to each team effectively responding to the four critical questions of a PLC?	• What collegial actions do you observe and monitor that contribute to each team effectively responding to the four critical questions of a PLC?
• How effective is your team communication?	• Which teams exemplify strong communication? Why? What are the contributing factors?	• Which teams exemplify strong communication? Why? What are the contributing factors?
• How does your team persevere with the daily challenges and demands of teaching and learning mathematics?	• Which teams demonstrate perseverance and a passion for their work? What evidence do you observe to show this?	• Which teams demonstrate perseverance and a passion for their work? What evidence do you monitor to show this?
• What is your team's current commitment level to the PLC process for mathematics?	• Which teams are fully committed to the PLC process? How would you describe that commitment?	• Which teams are fully committed to the PLC process? How would you describe that commitment?

Figure 1.1: Reflection tool—Coherence in your mathematics leadership practices.

*Visit **go.SolutionTree.com/MathematicsatWork** for a free reproducible version of this figure.*

Five Leadership Practices You lead with the following.	Description of Level 1	Requirements of the Indicator Not Present	Limited Requirements of This Indicator Are Present	Substantially Meets the Requirements of the Indicator	Fully Achieves the Requirements of the Indicator	Description of Level 4
Trusting Environment	Members of the mathematics team work individually and are passive participants during team meetings. Team members avoid conflict and are only willing to take ownership of their own students.	1	2	3	4	Each and every mathematics team member feels valued and that his or her opinions are important. Team members are willing to engage in action research, take risks, and consistently learn from each other. Team members are accountable to each other and address conflict as needed.
Relational Intelligence	Team members are complacent with each other and do not respect their colleagues during collaborative team time. Team members are easily defensive, do not listen to one another, and often betray private conversations.	1	2	3	4	Mutual respect is evident during collaborative team time. The strong passion for meaningful mathematics is observable in action steps as the team responds to the four critical questions of a PLC. Team members listen to one another and adhere to agreements.
Effective Communication	Team members often interrupt each other, or only limited voices speak during collaboration time. Agreements on the teams' work are not clear.	1	2	3	4	Team members listen to each other and ask clarifying questions to understand their colleagues' thinking. Team members understand decisions as agreements that each individual follows for action in the classroom.
Passion and Persistence	Team members display frustration and delusion and possibly disengage from team action because they feel the work will not make a difference. Team members shut down at the first indication of failure.	1	2	3	4	Team members are diligent in their work and have a "Do whatever it takes" attitude toward meeting their SMART goal and vision. Team members learn from mistakes, and they embrace any failures as a way to improve the work of the team.
Commitment to the PLC Process	Team members do not have clear expectations for actions that support the PLC process or are able to disengage from the PLC process.	1	2	3	4	Team members are committed to each other and the work they complete to ensure success for each and every learner. Team members identify the core values for their team mathematics work and take action on the expected work.

Figure 1.2: Discussion tool—Leadership practices rubric.

Visit go.SolutionTree.com/MathematicsatWork for a free reproducible version of this figure.

Author and speaker on leadership Robert Townsend (1970) states, "True leadership must be for the benefit of the followers, not the enrichment of the leaders." Servant leadership, a term that Robert K. Greenleaf (as cited in Frick & Spears, 1996) coined, requires making connections with those you are leading and fostering strong relationships to overcome challenges that may occur when addressing change.

The five leadership practices described in figure 1.2 intertwine and when you use them together, you can help effectively lead and support the collaborative PLC process for mathematics in your school or district. A solid foundation for supporting effective mathematics collaboration begins with the creation of a trusting environment. The sections that follow highlight each of the five leadership practices.

Trusting Environment

Trust is a key factor and essential for the effective risk taking your mathematics collaborative teams will require as they engage in action research (Kanold, 2017). Team members are more willing to engage in collaborative team actions when they feel that trust is firmly established and honored. Think of a time when you needed to trust your colleagues. What actions occurred before the process to provide you with the ability to trust and engage? Then read the story from author Matt Larson.

Personal Story **MATTHEW R. LARSON**

When my school district began to implement collaborative learning teams, one of the first steps during collaboration was that teacher teams needed to determine and make sense of the essential learning standards by unit. The process made a distinction between the *essential learning standards*—the overarching big mathematical ideas—and the individual *lesson learning targets*—targets that support the essential standards.

Some principals and teachers simply wanted the district office to provide a list of the essential learning standards by unit. Principals and teachers thought the list should be the same across the district and that teachers should spend their time collaboratively planning instruction instead of identifying the essential learning standards.

Efficiency and consistency were reasons for the district to provide the list, but it was critical from my perspective for teachers to engage in the discussions around the identification of standards. Only by engaging in the process of determining what the essential standards are could teachers truly make sense of the standards for themselves ("own the standards"), and, more importantly, reach agreement on a common expectation level and the mathematical practices students would engage in as they learned the standards. I finally convinced the majority of leaders that the process was as important as the product, but I had to trust that the process would lead to relatively common essential standards across the district.

The process worked! While there were some minor differences in wording, individual schools developed nearly identical lists of essential standards. More gratifying was the fact that numerous principals and teachers commented that the process itself was worthwhile and the conversations teachers had were invaluable as the work was foundational for the rest of the year's collaborative discussions.

When you are leading mathematics teams, the teams need to trust the collaborative team process. Establishing relationships is critical when developing the trust effective collaborative teams require.

Read the grade 1 team scenario in figure 1.3, and reflect on the actions of the team leader to create a trusting culture among team members.

Grade 1 Team Mathematics Scenario

At the start of the school year, the grade 1 team meets to start planning for the school year. The team leader had previously worked with the new teacher on the team during the summer and knew that he was nervous about the first day of school. The new team member has done his student teaching in grade 4 and is nervous about dealing with first-grade students.

The team leader starts the meeting with introductions. Then, he asks each returning team member to share the biggest mistake he or she has made when first teaching mathematics to grade 1 students. One teacher describes how she had assumed students could write their work on a whiteboard and share their thinking at the carpet. (She had come from grade 5.) Instead, she quickly realized she was going to have to teach students to show their work and explain it.

Figure 1.3: Grade 1 team mathematics scenario.

In the grade 1 scenario in figure 1.3, the team leader listens closely to the new team member and correctly reads the trust of the rest of the team. He knows that the other team members will be willing to support the new member by providing insight and stories to make the new member feel at ease. The mathematics team leader also poses the question in a way to suggest mistakes will happen. It is okay to laugh and learn from them in an effort to support risk taking.

The following are recommendations for developing a culture of trust.

- Ensure each and every team member feels valued and his or her opinions are important.

- Encourage risk taking and provide space for making mistakes.

- Foster building strong relationships by getting to know each team member and the strengths he or she brings to the team.

- Lead by example; treat others as you would like to be treated.

- Model vulnerability by taking risks and admit when you need support.

Trust is not something that always grows with time. You need to work to ensure the language and actions you and team members use in meetings help develop trust. This requires a second leadership practice, relational intelligence.

LEADER *Reflection*

In the grade 1 scenario (figure 1.3), what actions did the team leader take to make the new member feel at ease? What level of trust in the team did he establish by asking others to share their mistakes?

_____

_____

_____

_____

_____

_____

_____

_____

_____

_____

_____

_____

Personal Story 66 **BILL BARNES**

During my work with one school, I had the pleasure of observing a middle school mathematics coach leading her grade 7 team through the design of a high-quality common assessment. I was struck by the efficiency of the team as its members worked to professionally scrutinize each other's work. They were operating at a high level of trust and relational intelligence.

The following are the actions from the coach that I observed.

- The coach, an incredible listener, used paraphrasing and restating to clarify the meaning of her team members' contributions.

- The coach truly valued the input of each team member. She understood that building capacity was a process that took time.

- She clearly knew each team member's strengths and was able to draw on those strengths to elevate or revise their ideas.

- She built high team trust by investing in her team members' ideas, even when those ideas did not yet align to intended outcomes. She did not dismiss any ideas.

As a result of the team leader's commitment to building relational team intelligence, the team was functioning at a high level and members were learning and growing at an incredibly fast rate.

Relational Intelligence

As described in author Bill Barnes's story, your relational intelligence, or emotional intelligence, is an important leadership practice. *Relational intelligence* is your ability to understand, learn from, and comprehend the interpersonal dynamics of a collaborative team. The more relationally intelligent a leader is, the more he or she demonstrates mutual respect to team members (Saccone, 2009).

The first-grade team leader in the scenario in figure 1.3 displays strong relational intelligence as he listens to his new team member, understands the concerns, and identifies a way to make the new member feel at ease by listening to veteran teachers share their stories. How do you establish such a culture of caring as described in the story from author Bill Barnes?

So how do you develop relational skills? Several thought leaders provide insights on how to become a relationally stronger leader. Specifically, Daniel Goleman (2011) and Douglas Reeves (2006) share that there are specific practices you can use to create positive relationships within your collaborative teams.

- Be fully attentive in conversations. Listen without interrupting when team members are speaking instead of thinking about what you are going to say next.

- Be aware of the verbal and nonverbal responses and how they impact the team.

- Create an environment in which it just feels good to be with each team member. Exhibit genuine passion for each and every team member.

- Ask intentional questions to seek first to understand a team member's meaning.

- Never betray a private conversation.

These practices are not always easy to employ and may take intentional focus when you are working with teams so you can best model what behavior you expect of teammates as well. Use the leader reflection on page 18 to consider your current strategies for developing strong relationships with your teams and team members.

How do you currently demonstrate relational intelligence with the teams you lead based on the five bulleted criteria listed on page 17?

_____

_____

_____

_____

_____

_____

_____

_____

_____

Reflect on the five dysfunctions of a team—(1) absence of trust, (2) fear of conflict, (3) lack of commitment, (4) avoidance of accountability, and (5) inattention to results. The grade 8 team from the scenario in figure 1.4 displays a fear of conflict and avoidance of accountability.

What types of feedback would you offer to the team leader in order to provide support in this grade 8 team scenario?

_____

_____

_____

_____

_____

_____

_____

To build strong relational intelligence, you will also need to cultivate skills for effective communication.

Effective Communication

Communication isn't just about the conversations you engage in; it is the ability to listen without judgment to each and every team member. Strong listening skills empower teams to trust the collaborative process.

In learning to listen, you pay attention to both verbal and nonverbal communication that builds and maintains positive relationships. Read through the grade 8 team scenario in figure 1.4 and reflect on the listening feedback you would provide to the team leader or coach.

Imagine if the team leader in the grade 8 scenario instead prompts a dialogue about the data, saying "As you review these data, compare your individual results to the team results. With one partner, discuss possible causes for the gaps. Be prepared to share in a few minutes what you and your partner discover so we can come up with action steps as a team to address the causes." This prompt is one example of how you can create a safer environment with one-on-one conversations to effectively address possible fear of conflict and accountability avoidance.

Grade 8 Mathematics Team Scenario

The grade 8 mathematics team meets at the end of the first grading period with the instructional coach to engage in a data dialogue about the team's progress in meeting its SMART goal related to increasing the number of students passing in the first grading period for the course.

The teachers had submitted their overall grades for each class so that the team leader could compile all of the scores for the team to review. The leader displays these data at the beginning of the meeting, and the teachers quickly notice that one team member has a higher percentage of students failing compared to the rest of the team.

There is some eye rolling or looking down and away from these data, yet no team member specifically brings up the inconsistency when the team is debriefing during the data dialogue.

The team leader and the coach lead the team by asking about the trends in student performance and discussing ideas for possible re-engagement on each essential learning standard.

Not once does the team leader or other team members address the inequity in the grade distribution between the teachers, and how the team members can help each other for the next grading period.

Figure 1.4: Grade 8 mathematics team scenario.

When you encourage equal participation, you promote honoring the voice of every team member. Your leadership supports building a trusting environment, and in turn, holds every team member accountable for adding to the team decision-making process. This scenario also models the importance of developing strong communication skills and asking purposeful questions as you lead the mathematics team.

Part of your professional responsibility as a team leader is to facilitate learning. This requires you to *practice* listening to conversations and knowing questions to ask in order to further the thinking of each teacher on the team. You create space for all voices to be heard.

How can you improve your listening skills?

Karla Reiss (2007) suggests that well-crafted questions empower teachers to engage in powerful reflection and promote engagement in collaborative problem solving and creating self-efficacy simultaneously.

Consider the examples of questions and prompts in figure 1.5 and discuss the following.

1. Which questions are more effective at furthering team reflection?

2. Which questions can you strengthen with revisions?

As you can see from figure 1.5, not all questions are created equally. Which questions in figure 1.5 make teachers want to dig into the work? Which questions feel accusatory, judgmental, or make them uncomfortable?

Your questioning becomes a natural part of leading other adults. As such, you need to reflect on your own listening and questioning skills and examine whether they engage all team members.

How do the questions you ask during a team meeting promote deeper thinking? When you ask questions, how do you give everyone on the team a voice in responding to the questions? The quality of your questions will predict the impact of the discussion and thought process of your team members.

As with questions you would ask your students, a high-quality question for your mathematics team is open ended, requires more than a one-word response, and is nonjudgmental. Always assume positive intent when working with others.

Directions: Read the following questions. Identify the teacher thinking and team learning resulting from each one. Which of the questions should you modify to promote deeper thinking by the teacher team members?

1. As we review this assessment, what assessment questions do you agree with and why? What assessment questions do you disagree with and why?

2. This assessment is too long. What questions will we delete?

3. We have 1.5 hours to meet today, so what would success at the end of the meeting look like?

4. Do you have any questions about the agenda?

5. Looking at the student work we brought, what mathematics do the students most understand?

6. We all gave the same mathematical task, and my students were not able to complete the task. What is the problem with the mathematical task?

7. We each have a few students struggling with this mathematics concept. Can you each tell me how you taught it? Then, we can figure out the best way to re-engage the students.

8. We currently have fifteen students who are not meeting proficiency on _____ concept. We already tried manipulatives last year, and it didn't work. What do we do now?

9. What evidence will we collect to evaluate the effectiveness of the lesson we created?

10. This lesson did not work well. Who made this lesson?

Figure 1.5: Leader and team discussion tool—Sample team questions and prompts.

Visit go.SolutionTree.com/MathematicsatWork for a free reproducible version of this figure.

Your questions should relay your feelings toward various team actions. For example, "I know you have been working on . . . What has worked so far?, and let's brainstorm some other ideas to address . . ."

The following actions can improve your teams' communication skills.

- Record a team meeting session. Watch the session individually or with other leaders and take note of two things: (1) the types of questions being asked and who is asking and answering the questions and (2) who is engaged in the discussion of the questions.

- Keep a journal during a month of team meetings and reflect on the questions that you use to prompt your team and when those prompts result in productive meetings.

- Meet once a month with other team leaders and discuss questions that you plan to use to support a team's growth.

As you continue to foster the practices of effective communication and focus on relational intelligence to develop trust, you begin to explore the fourth leadership practice: the ability to persevere and exemplify your passion for mathematics and the work of leading and serving others—your ability to be passionate and persistent in your pursuit of equity.

Passion and Persistence

Leading mathematics teams and teachers through a collaborative process is not easy. It takes a strong passion for your work and the ability to persevere. Richard DuFour and Robert J. Marzano (2011) state it well: "The best educational leaders are in love—in love with the work they do, with the purpose their work serves, and with the people they lead and serve" (p. 194).

Psychologist and science author Angela Lee Duckworth (2016) calls this combination of passion and perseverance *grit*. The grittier, the better! *Grit* is the ability to persevere through the challenges that leading others will bring based on a strong passion for the work. In the context of behavior, grit is "firmness of character; indomitable spirit" ("grit", n.d.). Being *gritty* allows you to stay engaged in your leadership role and promotes *stick-with-it-ness*.

Grit is also about courage—the courage to develop trust when trust is not present, the courage to have a conversation with a peer who is breaking the norms the team members have agreed on, or the courage to establish strong structures, use effective strategies, and monitor the team's progress.

Grit means you will continue to try new strategies and redefine the goals to support the vision for you and your mathematics teams' continual growth, as described by author Mona Toncheff.

Personal Story   MONA TONCHEFF

In my first few years as a mathematics specialist, I focused on getting a district of more than 225 mathematics teachers to do two things: (1) come together to create a vision for mathematics teaching and learning and (2) bring the vision to life. There were bright spots on the journey as well as moments of frustration, confusion, and even times I questioned why I chose to leave the safety of the classroom.

However, even in the midst of challenges, I diligently moved us closer each and every year to meeting our vision for quality mathematics teaching and learning. The mathematics leadership team I supported created the vision and supporting action steps to keep us focused. On productive or difficult days, my actions and the actions of the team aligned to our agreed-on core value: *We believe all students can learn mathematics.*

This core value was the passion for my work. How could I support equitable learning experiences and guarantee access to meaningful mathematics for the twenty-seven thousand students we served if I didn't stay focused on completing the next task, organizing the upcoming professional development, planning for the next meeting, or focusing on the various other tasks of a mathematics leader?

Adam Grant, an author and professor at the Wharton School of the University of Pennsylvania, states that a critical skill for success is resiliency (as cited in Clifford, 2017). His research finds that strong resilience promotes your response to the challenges that arise in your leadership role (as cited in Clifford, 2017). Grant states "when you encounter a difficulty, a hardship, a challenge, how quickly and how effectively are you able to marshal strength and either overcome that challenge or persevere in the face of it?" (as cited in Clifford, 2017).

Use the leader reflection to reflect on actions you have taken that model your passion and persistence for leading mathematics.

LEADER *Reflection*

Consider your current leadership experiences, thinking about your actions and specific evidence that exemplifies passion and persistence. Provide an example that, for you, required persistence and resulted in a success with one of your teams.

_____
_____
_____
_____
_____
_____
_____
_____
_____
_____
_____
_____
_____
_____

As you reflect on your evidence that exemplifies grit and persistence, look for additional opportunities to display your passion and courage for mathematics.

The last leadership practice you will need to develop is to understand your teams' core values for mathematics instruction and assessment and how these core values support a commitment to the collaborative process.

Commitment to the PLC Process

You and your team members' core values permeate the work you do as professionals. They also impact the actions of each team member and his or her commitment to the PLC process.

As you engage in effective collaboration and commit to collectively responding to student learning, you teach and develop clarity about the core values of a PLC. How do you engage teams in conversations and reflection to identify core values and your collective commitments to the PLC process?

Richard DuFour and Robert Eaker (1998) state:

> What separates a learning community from an ordinary school is its collective commitment to guiding principles that articulate what the people in the school believe and what they seek to create. Furthermore, these guiding principles are not just articulated by those in positions of leadership; even more important, they are embedded in the hearts and minds of people throughout the school. (p. 25)

Core values stem from individual beliefs. When beliefs align, teams can flourish; when beliefs do not, teams can experience deep conflict and frustration (Hirsh & Killion, 2007).

To engage your team in uncovering the team's core values, you ask each team member to reflect on his or her core values and reach consensus on the team's collective commitments. (You explore collective commitments in detail in chapter 2.)

Figure 1.6 (page 22) is a list of words and phrases that represent possible individual values (Aguilar, 2013). Before using the list with your team, read through the list yourself and consider if you might add something to the list.

Once you and your team members identify your core values, ask each team member to share his or her top-three core values and consider how those individual core values support team actions. Come to agreement on the team's core values and how the values will support your professional work. These core values also inform the collective commitments that your team will create.

Other considerations are the productive and unproductive beliefs about mathematics teaching and learning of your team members.

Directions:
1. Read through the following values, and circle ten values that are important to you.
2. Cross off five of those values, leaving the five that are most important to you.
3. From the list of five values, cross off two, leaving the three values that are most important to you. Your core values are most likely a subset from your final three selections.

Acceptance	Effectiveness	Humor	Productivity
Achievement	Efficiency	Imagination	Recognition
Adventure	Equality	Independence	Reflection
Affection	Equity	Influence	Religion
Altruism	Excellence	Initiative	Reputation
Ambition	Excitement	Integrity	Respect
Appreciation	Expertise	Interdependence	Responsibility
Arts	Espression (or self-expression)	Intuition	Results
Authenticity	Fairness	Joy	Risk taking
Authority	Faith	Justice	Romance (or self-romance)
Autonomy	Fame	Kindness	Service
Balance	Family	Knowledge	Sharing
Beauty	Flexibility	Leadership	Solitude
Belonging	Focus	Loyalty	Spirituality
Caring	Forgiveness	Making a difference	Success
Celebration	Freedom	Meaningful work	Support
Challenge	Friendship	Mindfulness	Teamwork
Choice	Fun	Nature	Time
Collaboration	Goals	Nurturing	Togetherness
Commitment	Gratitude	Order	Tolerance
Communication	Growth	Passion	Tradition
Community	Happiness	Peace	Travel
Compassion	Health	Personal development	Trust
Connection	Helping others	Personal growth	Truth
Contribution	High expectations	Perseverance	Unity
Cooperation	Honesty	Pleasure	Variety
Creativity	Hope	Positive attitude	Zest
Democracy	Humility	Pride	

Source: Aguilar, 2013.

Figure 1.6: Team discussion tool—Exploring core values teacher team tool.

*Visit **go.SolutionTree.com/MathematicsatWork** for a free reproducible version of this figure.*

Teaching is a cultural activity, and our cultural beliefs about the teaching and learning of mathematics can create a barrier to the implementation of effective teaching practices (National Council of Teachers of Mathematics [NCTM], 2014). What is your belief about effective mathematics instruction?

Your beliefs about mathematics instruction derive, in part, from your background; nearly everyone has experienced mathematics education. In fact, most North American high school graduates have experienced about 1,500 hours of mathematics instruction (Larson & Kanold, 2016). This experience creates a powerful cultural expectation for mathematics teaching and learning.

There are still many educators, not to mention parents, who believe that students should be taught mathematics as they were taught, with an emphasis on memorizing facts and procedures and repetitive practice. In contrast, research indicates that effective instruction

Personal Story 66 **M O N A   T O N C H E F F**

When working with a newly formed group of district-level mathematics coaches, we noted that one of their challenges was getting their mathematics teams focused on engaging in the PLC process. Some of the actions or support for the teams I observed while working with them varied from coach to coach. I chose to use the Exploring Core Values Teacher Team Tool activity (figure 1.6) to address understanding each other's values and how those values impact their work. As the coaches shared their final three selections, they were surprised at the range of differences in their choices. The resulting conversations generated an understanding of one another's core values and how they could work together to draw on each other's strengths to better support the mathematics teams in the district.

engages students in challenging tasks, supports students' productive struggle, and encourages mathematical discourse and the development of positive student mathematical identities (Smith, Steele, & Raith, 2017). The goal of effective mathematics instruction is to ensure that each and every student succeeds in doing high-quality academic work rather than merely reproducing memorized procedures with speed and accuracy.

Examining the table of unproductive and productive beliefs about teaching and learning mathematics in NCTM's (2014) publication *Principles to Actions* is a good team discussion tool that will reveal your mathematics teams' core beliefs concerning effective mathematics instruction. As you read and respond together to the beliefs in figure 1.7, consider your team's actions and

Beliefs About Teaching and Learning Mathematics	
Unproductive Beliefs	**Productive Beliefs**
Mathematics learning should focus on practicing procedures and memorizing basic number combinations.	Mathematics learning should focus on developing understanding of concepts and procedures through problem solving, reasoning, and discourse.
Students need only to learn and use the same standard computational algorithms and the same prescribed methods to solve algebraic problems.	All students need to have a range of strategies and approaches from which to choose when solving problems, including, but not limited to, general methods, standard algorithms, and procedures.
Students can learn to apply mathematics only after they master the basic skills.	Students can learn mathematics through exploring and solving contextual and mathematical problems.
The role of the teacher is to tell students exactly what definitions, formulas, and rules they should know and demonstrate how to use this information to solve mathematics problems.	The teacher's role is to engage students in tasks that promote reasoning and problem solving and facilitate discourse that moves students toward shared understanding of mathematics.
The role of the student is to memorize information that the teacher presents and then use it to solve routine problems on homework, quizzes, and tests.	The student's role is to be actively involved in making sense of mathematics tasks by using varied strategies and representations, justifying solutions, making connections to prior knowledge or familiar contexts and experiences, and considering the reasoning of others.
An effective teacher makes the mathematics easy for students by guiding them step by step through problem solving to ensure that they are not frustrated or confused.	An effective teacher provides students with appropriate challenge, encourages perseverance in solving problems, and supports productive struggle in learning mathematics.

Source: NCTM, 2014.

Figure 1.7: Team discussion tool—Productive and unproductive beliefs about teaching and learning mathematics.

*Visit **go.SolutionTree.com/MathematicsatWork** for a free reproducible version of this figure.*

conversations about mathematics instruction. Which beliefs align with your current instructional practices?

You can lead your teams through a reflection about their beliefs using figure 1.7 (page 23) to help better understand *each team member's* current beliefs. When teams understand the connections between productive beliefs and actions that promote learning for all, you and your teams are able to align team actions to support more productive beliefs about student learning.

In addition, you can use NCTM's *Principles to Actions* Professional Learning Toolkit (www.nctm.org /ptatoolkit) to support your teams' implementation of effective instructional strategies and examination of their beliefs. (Visit **go.SolutionTree.com/Mathematics atWork** to access live links to the websites mentioned in this book.)

Reflect, Refine, and Act on the Five Leadership Practices

The five leadership practices when combined support the development of a culture of caring and strong team relationships. Imagine the power of your teams when

each individual member feels honored and respected and is willing to complete any task that supports the team's purpose.

When you utilize and develop these personal leadership practices, your mathematics teams and their members will feel valued and better understand the common purpose of the team. Team members trust the process, and they are willing to take risks. They are passionate and persevere as they collaborate to support the overall purpose and commit to the energy and work that they need to collectively respond to student learning and mathematical understanding.

Use the final leader reflection to consider once again your growth and development around these five personal leadership practices in this chapter.

It is your professional responsibility to develop and foster strong relationships and engage your mathematics teams toward effective collaboration. Part of the professional work of leading others is to intentionally engage in the leadership practices for this purpose and then choose *strategies* that require your teams to focus on the right work, as described next in chapter 2.

LEADER *Reflection*

Which of the five leadership practices do you consider a personal strength?

How can you become more engaged in a leadership practice in which you might not be as strong as the others?

_____

_____

_____

_____

LEADER RECOMMENDATIONS

Leadership Practices

- Create a culture of trust so teams are willing to take risks, try new practices, and learn from each other.

- Continue to grow emotional and relational intelligence to understand, learn from, and comprehend the interpersonal dynamics of your collaborative teams.

- Take time to listen to your team members without judgment. Continue to use

effective communication to build trust and empower teams.

- Persevere through the challenges that may occur when leading others based on your strong passion for the work.

- Understand your team core values to ensure team actions align with productive beliefs for mathematics teaching and learning.

CHAPTER 2

Five Leadership Strategies for Effective Collaboration in Mathematics

Leaders enhance the effectiveness of others when they provide clarity regarding what needs to be done and ongoing support to help staff succeed.

—*Richard DuFour and Robert J. Marzano*

As you support your mathematics collaborative teams, the first order of business is to establish a common purpose and create clarity for the team expectations, outcomes, actions, and behaviors.

- Do all team members understand why collaboration is important?
- Can all team members describe effective collaboration?
- How do you monitor and support effective collaboration?

If there are mathematics team leaders, mathematics coaches, and administrators in your school, complete your responses to figure 2.1 (page 26) together.

Once you have completed the mathematics leadership reflection activity in figure 2.1, you can use the leadership strategies scoring rubric in figure 2.2 (page 27) to rate each team you lead and coach in the five strategic areas:

1. Common purpose
2. Collective commitments that focus on collaboration

3. Evidence of success
4. Continuous reflection and refinement
5. Mutual accountability to the professional work of the team

In which of the five strategic leadership areas are your teams strongest? Which challenge them? These five specific leadership strategies ensure your mathematics teams are collectively responding to the four critical questions of a PLC (DuFour et al., 2016):

1. What do we want all students to know and be able to do?
2. How will we know if they learn it?
3. How will we respond when some students do not learn?
4. How will we extend the learning for students who are already proficient?

Keeping your eye on your ratings from figure 2.2, read the personal story from author Mona Toncheff that follows on page 28. Decide how you might rate the situation in her story against the five criteria from figure 2.2.

Team Leader	Mathematics Coach	Site-Level Leader
• What is your team's common purpose for its professional work? How do you and your team members work toward this common purpose for learning mathematics?	• Which mathematics teams have a clear sense of purpose? How does each team member articulate and own the purpose and share it with you?	• Which mathematics teams have a clear sense of purpose? How does each team member articulate and own the purpose and then report it to you?
• How do you and your team members hold each other accountable to your professional work as teachers of mathematics?	• Which mathematics teams support each other and adhere to the norms they have established? How do teams monitor and support the norms? How do you know?	• Which mathematics teams support each other and adhere to the norms they have established? How do teams monitor and support the norms? How do you know?
• What are your team's mathematics goals, and how does your team collectively work toward meeting those goals?	• Which mathematics teams use goals to drive their work? What successes have you observed with goal-oriented teams? How do you support the team goals?	• Which mathematics teams use goals to drive their work? What successes have you observed with goal-oriented teams? How do you monitor and support the team goals?
• How do you and your team engage in continual reflection and learning to increase student achievement in mathematics?	• Which mathematics teams demonstrate a continuous cycle of inquiry and learning and why? How do you know?	• Which mathematics teams demonstrate a continuous cycle of inquiry and learning and why? How do you know?
• How do you and your team members hold each other accountable for the products the team creates and the expectations of quality mathematics lessons, assessments, and interventions?	• Which mathematics teams equally share the responsibility of the teamwork? How do you know the team members hold each other accountable?	• Which mathematics teams equally share the responsibility of the teamwork? How do you know the team members hold each other accountable?

Figure 2.1: Reflection tool—Leadership strategies.

*Visit **go.SolutionTree.com/MathematicsatWork** for a free reproducible version of this figure.*

Leadership Strategies	Description of Level 1	Requirements of the Indicator Are Not Present	Limited Requirements of This Indicator Are Present	Substantially Meets the Requirements of the Indicator	Fully Achieves the Requirements of the Indicator	Description of Level 4
Common Purpose	Team members work in isolation improving their individual student performance.	1	2	3	4	The clearly articulated mathematics teaching and learning vision is the driving force for the team, and each and every team member has a deep understanding of how his or her actions and the team actions align with the vision.
Collective Commitments That Focus on Collaboration	Team members do not respect the norms, or there are no clear expectations for team behaviors. Accountability is not a norm for the team.	1	2	3	4	Team members adhere to the established norms and hold each other accountable for their actions. Team member actions align to their values as a team. The team consistently evaluates the norms and makes changes as it needs to.
Evidence of Success	Team members may have a SMART goal; however, the team does not evaluate student learning throughout the year to monitor progress on meeting the goal. Team members lack the coherence in actions they need to meet the goal.	1	2	3	4	The team has clearly articulated SMART goals (long term and short term) and supportive action steps. The team has a shared knowledge of how the action steps will support meeting the goals. The goals and action steps align to the mission and vision of the school or district.
Continuous Reflection and Refinement	Team members rely on current teaching practices and rarely share instructional strategies. Team members focus on creating common artifacts, and do not use evidence of student learning to make adaptations to instruction.	1	2	3	4	Teams engage in continual reflection and fully engage in protocols for effective teacher team action. Protocols ensure all members predict, analyze, problem solve, and create actions based on the outcomes of the discussions. Protocols teams use create team transparency.
Mutual Accountability to the Professional Work of the Team	The accountability for the team's success is solely on the team leader. There are some shared artifacts, yet team members do not collectively share resources.	1	2	3	4	Team members and the team leader collectively share the responsibilities of and accountability for improving the mathematics teaching and learning. All team members share agendas, minutes, and artifacts, and there are clear expectations for the quality of the collaborative teamwork.

Figure 2.2: Team discussion and reflection tool—Leadership strategies scoring rubric.

Visit go.SolutionTree.com/MathematicsatWork for a free reproducible version of this figure.

Personal Story 66 MONA TONCHEFF

I still remember my first collaborative team meeting and the feeling of accomplishment I had at the end of the meeting. I had survived! I was the designated algebra 1 team leader. From our initial PLC training, I knew we needed to work on common homework and common assessments. What I did not know was how difficult it would be to determine anything common due to the differences on our team among the instructional practices and beliefs of members.

During our first collaborative team meeting, my team members' responses to scoring a common assessment surprised me. For example, when we were discussing the essential standard on solving equations, one teacher said she would only give students credit for a response if they showed all of the steps vertically. Another teacher said that he didn't care if there is evidence of solving the problem; he only looks at whether or not the answer is correct. None of the team members were willing to compromise on the scoring of the assessment, and I was not sure they would even give the assessment the team designed that day.

The team in author Mona Toncheff's story had no established purpose and would have scored low for this criteria as listed in figure 2.2 (page 27). Its members focused on creating a common assessment rather than developing the rationale as to *why* the team needed to create the common assessment. When two of the members chose to give their own assessment at the end of the unit, they did not understand how their selfish actions were harmful to the team and would continue to create inequities in student learning. The team needed a discussion about the greater purpose of collaborative teams and common assessments. The team was not clear on its PLC foundation—its mission, vision, values, and goals.

The Four Pillars of a PLC

The four pillars of a PLC are collaboratively shared (1) mission, (2) vision, (3) values, and (4) goals (DuFour et al., 2016). When you serve mathematics collaborative teams as a leader or coach, you start by building a shared understanding of these four pillars and provide the rationale for the purpose of the team and team actions that align to these four pillars. By creating a shared mathematical teaching and learning mission and vision, you can work with your teams to establish collective commitments and monitor behavior as well as set short-term and long-term goals to meet the

mathematical teaching and learning vision (DuFour et al., 2006, 2010; Mattos et al., 2016). Examine figure 2.3. This working document provides space to answer questions that align to the four pillars of any PLC at Work team (DuFour et al., 2016).

The mission pillar for mathematics teachers is simple, and it is established in the title for this series: *Ensure each and every student can learn mathematics.* Engaging teams in reflecting on their productive beliefs (page 23) and helping your teams understand their common purpose—the *mission* for ensuring all students can learn mathematics—can help you to connect others to the fundamental purpose. When you know your why, your what (the vision of what you hope your mathematics program will become) has more meaning.

As you read through the grade 3 team scenario in figure 2.4 (page 30), consider qualities that demonstrate common direction (vision) and the actions you can take to support the team's work.

When reading through the scenario, you might think, "Wow, the team leader has created several accountability structures for effective collaboration." However, this scenario also exemplifies that strong relationships do not necessarily increase student achievement.

Mission Why do we exist? What is our purpose as teachers of mathematics?	Vision What kind of grade-level mathematics team or department do we hope to become?
Values What collective commitments must we make to move our mathematics team in the direction we want it to go in? How must we behave in order to make our vision a reality?	Goals What are the expected student learning outcomes of our work? What steps are we going to take to ensure those outcomes, and when will we take them to move toward our vision for mathematics?

Figure 2.3: Mission, vision, values, and goals team-planning document.

*Visit **go.SolutionTree.com/MathematicsatWork** for a free reproducible version of this figure.*

The team agrees on the essential learning standards and defines proficiency on each standard (a common vision for assessment and grading), and it agrees on how to use these data to re-engage students (a common vision for intervention); however, it fails to come together to create a common vision for mathematics instruction.

Effective collaborative teams will design a shared vision for mathematics assessment, instruction, grading, homework, and intervention to ensure that the team directs its work all toward the same purpose.

Using figure 2.3, work with your teams to create a vision statement around instruction. (Visit **go.SolutionTree .com/MathematicsatWork** for a vision-development-brainstorming activity.)

Figure 2.5 (page 30) is a sample mathematics vision statement with supporting action steps (core values) the grade 3 team could commit to and work toward.

Consider again the grade 3 team in figure 2.4. After the team establishes a vision for instruction and realizes

as part of the brainstorming process that student-to-student discourse is one of its challenges, its next action step is to be more transparent with classroom lessons and observe each other trying new instructional strategies, asking what works and what does not work so well.

In supporting the grade 3 team with this action step, you will need to provide the teachers with the release time needed to observe each other. You also need to coach them on how to honor their collective effort when they are observing each other and their students, as teachers need to behave as respectful learners too.

Change does not occur overnight; you will need to work with each team on the mathematics instruction, grading, or homework, assessment, and intervention shifts that should occur each and every year as part of a three- to five-year plan to support the complex challenges of the teams' work.

Grade 3 Mathematics Team Scenario for Common Purpose

A grade 3 team has been working as a collaborative team for six years. The team has a shared Google Drive folder where it stores its common artifacts. The team leader has a specific folder for the agendas for each meeting and posts minutes for the instructional coach and principal to review each week.

On the minutes, team members check off a box each week addressing which critical question of a PLC they are answering. The team has a strong culture of trust and is willing to compromise and ask questions about instruction. The team analyzes the common assessments and creates small-group instruction plans on essential standards and concepts that students are struggling with during the unit.

At the end of the six years, the state assessment scores for grade 3 mathematics show little to no growth in student achievement. The team leader is frustrated because she feels like the work the team is doing is not making a difference.

When the instructional coach asks her to identify the root cause, the team leader feels that the team's conversations about instruction are lacking. At the next team meeting, the coach asks every team member to describe specific teacher actions and strategies that impact student learning and if the team agrees on these intentional teacher actions for mathematics.

The team members have a variety of responses that show there is little coherence with their instructional expectations.

After the team meeting, the team leader tells the coach the team has never explicitly stated what good mathematics instruction looks like or what it might sound like in the classroom.

Figure 2.4: Reflection tool—Grade 3 mathematics team scenario for common purpose.

LEADER *Reflection*

Reread the grade 3 scenario from figure 2.4. Regardless of the grade-level teams or mathematics course teams you lead, how does the scenario compare to your team's current progress regarding the purpose of mathematics instruction?

_____

_____

_____

_____

_____

_____

_____

_____

Team, Department, or School Vision for Mathematics Instruction

We want to create student-centered classrooms where mathematics and sound student reasoning are the norm.

Three action steps:

1. We will seat students in teams of four and post sentence starters on each set of four desks to model how students can use probing questions when working with their teams.

2. Once a month, we will engage in instructional rounds and teacher observation with colleagues to provide feedback on the transition from teacher-led discussion to student-led discussion.

3. We will create a problem-solving form that we will consistently use to model how students use mathematics and student reasoning as the final authority when critiquing and justifying their reasoning.

Figure 2.5: Sample mathematics vision statement with action steps.

TEAM RECOMMENDATIONS

Establish a Common Purpose

With your team, consider the following recommendations.

• Once your team clearly articulates a vision for mathematics instruction, assessment, and intervention, use the vision to guide the work of the team.

• Ask team members to work together to align specific team actions to the agreed-on vision for mathematics.

• Create three- to five-year plans to support meeting the vision using specific team actions each year to move toward the vision.

A compelling vision provides a foundation and rationale for the team's work but does not necessarily define for your teachers on the team how to effectively work together to develop their collective commitment toward collaboration.

Values: Collective Commitments That Focus on Collaboration

Mathematics collaborative teams and leaders need to have agreed-on commitments describing how the team members are going to work together toward meeting the vision. *Collective commitments* are the promises team members make and honor in order to meet their shared vision for teaching and learning mathematics.

What if a strong component of your instructional vision is effective mathematical student-to-student discourse, but every classroom in the building is set up with desks in traditional rows? How do the actions of team members align to the vision? If the team wants to meet the instructional vision, it will also have to commit to changing the seating arrangements in the classroom to foster student-to-student discourse.

Examine figure 2.3 (page 29) and the questions your team considers when creating commitments. For example, if you value student-to-student discourse and understand the power of students sharing their mathematical thinking, then the team members write a commitment tied to their vision for mathematics instruction.

Clarifying these collective commitments is "one of the most important and, regrettably, least utilized strategies in building the foundation of a PLC" (DuFour et al., 2016, p. 42). The following are some questions for you and your team to consider when establishing the collective commitments.

• How do we want our students to engage in a mathematics lesson (vision for instruction), and what actions will we need to take to support this?

• In relation to our vision for formative feedback and assessment processes, what commitments do we need to make to reach our assessment literacy vision?

• How do we want students to re-engage in and own their learning (our vision for intervention)?

• What type of feedback will we provide to students (vision for grading), and how will students and teachers utilize the feedback?

• What collective independent practice and feedback (vision for homework) will we provide, and how will we require students to take action?

Figure 2.6 (page 32) shows a sample set of collective commitments from a grade 6 team. Read through the collective commitments, and think about the actions that align with the collective commitments.

The first commitment might require members to discuss their assessment practices and how they encourage students to re-engage in learning. The third commitment would require the team leader to schedule time with each unit to collaboratively score common assessments and come to agreements on the feedback students receive. Now use the leader reflection to take some time to reflect on your team's current collective commitments for improving the mathematics program in your school.

Grade 6 Team Collective Commitments

We will collaboratively:

- Provide students with multiple opportunities to demonstrate their proficiency

- Discuss the best instructional strategies, and evaluate the impact of these strategies on student learning

- Score common assessments to ensure our feedback is consistent from teacher to teacher on our team

- Plan the independent practice and classroom practice (homework) to provide rigor balance

Figure 2.6: Sample team collective commitments.

LEADER *Reflection*

Reread the grade 6 collective commitments from figure 2.6. What are your team's current collective commitments regarding mathematics instruction and assessment? What do you need to add or revise to ensure your team actions support your team's vision for the mathematics program?

_____
_____
_____
_____
_____
_____
_____
_____

Once the team clarifies its collective commitments, a next step is to establish *norms* on how the team interacts—public agreements on how team members will behave, interact, and support each other.

Google is well known for its collaborative work environment, and yet, within Google, some teams are more effective than others (Duhigg, 2016). Why? Google

hired researchers to find out. These researchers find that teams that had clear guidelines and expectations for their work established *psychological safety*—an environment that has empathy and encourages emotional conversations. Such safe environments were able to outperform other teams. The researchers also notice that two equally effective teams could have different expectations or norms for their interactions and still be effective. What matters is that the norms are personalized to each team to address the different relationships on each team (Duhigg, 2016).

For your norms to be effective and contribute to a culture of learning, they have to be clearly articulated and collaboratively established. Teams too often complete norms at the beginning of the year and never or seldom refer to them again: "We have our team norms—check; we are done." However, you need team norms to guide team culture and engagement.

Norms become more than a list of professional courtesies. Compare the two sets of norms that appear in figure 2.7. What are the similarities and differences?

As you compare and contrast the two sets of norms, consider the following norm-setting advice from DuFour et al. (2010).

- Each team should be responsible for creating its own norms, and each team member should be involved in creating the norms.

- Teams should state norms as commitments to act or behave in certain ways rather than reflect a list of core values.

- Team members should review norms consistently once they are created and then evaluate them formally at least twice throughout the school year.

- Instead of a long list of norms, teams should have a list of norms that focus on the behaviors that are impacting team effectiveness.

- Norms should always include an agreement on how teams will make decisions and how each team member is going to agree and support the decisions.

Team 1 Norms	Team 2 Norms
Norms for all teams at Lincoln Middle School:	**Grade 4 collaborative team norms:**

• We believe in a positive environment for the teachers and the students.
 	
• We will only use cell phones in case of an emergency.
 	
• Majority rules.
 	
• Come prepared to the meeting.
	• We will maintain a positive tone at our meetings.
	
	• We will not complain about a problem unless we can offer a solution.
	
	• We will begin and end our meetings on time and stay fully engaged throughout each meeting.
	
	• We will contribute equally to the workload of this team.
	
	• We will listen respectfully and consider all thoughts and opinions before making an informed decision.
	
	• We will hold one another accountable to the work.
	
	• We will use the *fist to five* strategy to come to consensus.
	
	   ◦ *Five fingers:* I love this proposal. I will champion it.
	
	   ◦ *Four fingers:* I strongly agree with the proposal.
	
	   ◦ *Three fingers:* The proposal is okay with me. I am willing to go along.
	
	   ◦ *Two fingers:* I have reservations and am not yet ready to support this proposal.
	
	   ◦ *One finger:* I am opposed to this proposal.
	
	   ◦ *Fist:* If I had the authority, I would veto this proposal, regardless of the will of the group.

Figure 2.7: Sample team norms.

LEADER *Reflection*

What norm-setting advice from DuFour et al. (2010) should the Lincoln Middle School principal in figure 2.7 adhere to regarding the norms for all teams? What could they learn from the grade 4 team's norms? What feedback would you provide the team to create more effective norms?

_____

_____

_____

_____

_____

_____

_____

_____

_____

_____

_____

_____

_____

_____

_____

Kenneth C. Williams (2010) states that there are three things that occur when you fail to address team members who are breaking the agreed-on norms for your team.

1. Unspoken tension and frustration grow within the collaborative team when you don't address the norm violation.

2. With no established protocol, reaction to the confrontation becomes defensive when you address the norm violation inappropriately.

3. Too early in the process, the team takes the issue to the principal for him or her to handle when you don't address the norm violation.

As part of establishing norms, you and your team establish a process for managing healthy conflict that arises. Teams create signals or simple reminders to use during team meetings as team members self-monitor behaviors that do not align with the agreed-on norms. If a team member is consistently violating the norms, you need to meet with the team member to understand the norms he or she is violating as you help him or her to re-engage with the agreed-on norms.

When attempts from team leaders, peers, or instructional coaches do not bring about required changes, it will be necessary to recruit assistance from an administrator who can address the team member's violations and ensure that he or she begins to adhere to the team norms (Mattos et al., 2016).

What norms did your collaborative teams agree on? Are your current norms tied to specific team actions that all team members honor? Use figure 2.8 with your team to strengthen your norms. Begin by asking each team member to answer the questions in the left column individually.

Once every team member completes figure 2.8, ask him or her to share one thought at a time until everyone gives input. Once you hear all voices, you and your team can agree on three to five norms that will support effective collaboration. Before ending the meeting, identify how the team will evaluate the effectiveness of the norms and if the team needs to create new norms to address a different adult action impeding effective collaboration. Visit **go.SolutionTree.com/MathematicsatWork** to see a completed example of figure 2.8.

Figure 2.9 is an example of a survey that you can use with your mathematics team to evaluate its norms throughout the school year.

Ask each mathematics team member to complete the survey individually and then, discuss the survey results as a team. Depending on the trust level of the team, you can discuss the survey together, or you, as the leader, can collect the anonymous results and use them for a team discussion. Teams can also use data from the survey to inform next steps and create or revise norms to ensure effective collaboration.

If ineffective adult actions are preventing effective collaboration or creating inequitable student learning across the team, utilizing norms is an effective strategy to build trust. When you and your teams adhere to the established norms and collective commitments, you will build a strong community of learners willing to learn from one another and engage in the six mathematics team actions the other three books in the *Every Student Can Learn Mathematics* series describe in detail.

TEAM RECOMMENDATIONS

Develop and Monitor Collective Commitments

- Create collective agreements that align to your mission and vision.

- Ensure you and team members adhere to the established norms and hold each other accountable for actions.

- When creating action steps, ensure alignment between core values, productive beliefs, and action steps.

- Work with team members to consistently evaluate the norms and make changes as you need to.

- Work collaboratively with team leaders, coaches, and site-level leaders to mediate ineffective adult actions.

The following section beginning on page 36 outlines the measure of success that will be the target you and your mathematics teams are collectively trying to achieve.

Guiding Questions for Team Norm Creation	Comments and Possible Norms
What do you expect of yourself if you are to be part of a great team?	
What do you expect of others as part of a great team?	
What type of feedback do you want from your colleagues in order to stay engaged in the collaborative work of the team?	
How should your team keep an open line of communication with one another?	
How should you come to agreement on a team decision to reach consensus?	
How should your team celebrate your teamwork and weekly effort?	

Figure 2.8: Team discussion tool—Guiding questions to develop norms.

*Visit **go.SolutionTree.com/MathematicsatWork** for a free reproducible version of this figure.*

Survey of Team Norms				
Team: _____ Date: _____				
	Strongly Disagree	**Disagree**	**Agree**	**Strongly Agree**
I know the norms my team has established.				
Members of my team honor the established norms.				
We have established our consensus norm, and it supports team decision making.				
Our team has clear expectations for monitoring norms.				
Our team has a protocol for addressing team members who are not following the norms.				

Source: Adapted from DuFour, DuFour, & Eaker, 2006.

Figure 2.9: Team discussion tool—Survey of team norms.

*Visit **go.SolutionTree.com/MathematicsatWork** for a free reproducible version of this figure.*

Goals: Evidence of Success

The third big idea of a PLC is a focus on results (DuFour et al., 2016). The best way to ensure that collaborative teams focus on results is to insist that each and every team establish both short-term and long-term evidence of success and collectively describe the actions that the team will work toward to achieve that success.

You and your mathematics team create SMART goals to clarify the evidence of success and articulate the action steps to meet your goals. SMART goals are (Conzemius & O'Neill, 2014):

- **Strategic and specific**—Specific and aligned to the site- and district-level goals, vision, and mission

- **Measurable**—Include the current reality of the measure of success and the desired percentage

- **Attainable**—Well-defined and achievable

- **Results oriented**—Tied to an agreed-on measure of success

- **Time bound**—Include a specified amount of time to measure success

Goals are the targets that beckon for collaborative teams. Goals answer the questions, "What steps are we going to take, and when will we take them to move toward our mission?"

When you lead a team focused on effective teacher collaboration and the team goal is to increase students' mathematical achievement, learning outcomes will become more equitable and students' learning experiences will cultivate positive mathematics identities and an increased sense of agency. The SMART goals and the supporting action steps you and your mathematics team provide are the anchor for all collaborative work.

Before you and your team create SMART goals, work with team members to come to agreement on the *measure of success* you will monitor to gauge student learning. Team members can look at multiple student-achievement measures including but not limited to:

- State (or province) and national assessment measures

- Local district assessment measures

- Site-level common assessment results for each mathematics unit of study

- Site-level mathematics grade data

- Program results (for example, an increase in the percentage of students in advanced placement mathematics that match the district's overall demographics, or a decrease in the percentage of students in below-level mathematics courses)

Refer to AllThingsPLC's Tools & Resources page (www.allthingsplc.info/tools-resources) for data-picture tools for identifying measures of success. (Visit **go.Solution Tree.com/MathematicsatWork** to access live links to the websites mentioned in this book.)

SMART goals can also be either long term or short term. For example, a grade 5 team wants to focus on the end-of-year state (or province) assessment for its measure of success, and it sets the following two goals.

1. **Long term:** We will increase the percentage of students proficient on the end-of-year state assessment from 67 percent to 75 percent.

2. **Short term:** We will increase the percentage of students proficient on the unit assessment on equivalent fractions from 55 percent to 70 percent.

Although the grade 5 team measure of success is the end-of-year assessment, the team includes a short-term goal for the equivalent fractions unit because students continue to struggle with the concepts. In the past, their students struggled to learn the essential standards for this unit, so the team (perhaps with your help) decides to focus its energy on digging deeper into each team members' own knowledge about how students develop an understanding of equivalence with fractions.

When teams set short-term goals for student learning, they can also celebrate short-term wins and monitor and adjust to ensure that they are on target to meet their longer-term goals.

Notice both of the grade 5 SMART goals have two data points—a *from point* and a *to point*. Once the team agrees on a measure of success, you will need to help team members gather the historical data necessary to define the current reality (the *from* point) of the collaborative team and then, work with the team to write SMART goals (the *to* point) and the accompanying action steps. Strategies for how to collect historical data trends appear in chapter 3 (page 61).

Use figure 2.10 as a tool to create a SMART goal with supporting action steps for the teams you support.

SMART Goal and Data-Analysis-Planning Protocol

Team: _____   Date: _____

Long-term SMART goal: _____

Short-term SMART goal: _____

Answer the following questions to draft a plan to monitor these goals.

1. What baseline data will you use as an initial measure of student learning related to the SMART goal?

2. Which instructional strategies will help accomplish these short- and long-term SMART goals?

3. When will you give common mid-unit and unit assessments to measure progress toward the SMART goal?

4. When will you analyze data from the common unit assessments?

5. What will we look for in student work as evidence that you reach the long-term goal?

6. What interventions or extensions are we planning to utilize to support students?

7. How will we celebrate when we reach the goal?

Figure 2.10: Team discussion tool—SMART Goal and Data-Analysis Planning Protocol.

*Visit **go.SolutionTree.com/MathematicsatWork** for a free reproducible version of this figure.*

Once you establish the SMART goal, work with your teams to consider the necessary action steps to meet the goal. *Action steps* are the activities that collaborative teams complete or artifacts they create to meet the SMART goal. Creating a goal that has a sense of urgency is important; however, the most important part of SMART goal setting is being explicit about the action steps teacher teams need to take to achieve the goal.

When writing action steps with teams, in addition to using figure 2.10 (page 37), ask each team member to reflect on the six team actions listed in figure 2.11 for possible action steps to support meeting the goal. Teams can also use the following rubrics from the *Every Student Can Learn Mathematics* series for specific steps to self-assess within each team action (see Kanold, Barnes, et al., 2018; Kanold, Kanold-McIntyre, et al., 2018; Kanold, Schuhl, et al., 2018).

- *Mathematics Assessment and Intervention in a PLC at Work*

  - Figure 1.1: Team Discussion Tool—Common Assessment Self-Reflection Protocol

  - Figure 2.1: Team Discussion Tool—Assessment Instrument Quality Evaluation Rubric

  - Figure 5.2: Mathematics Intervention Program Evaluation Tool

- *Mathematics Instruction and Tasks in a PLC at Work*

  - Figure P1.2: Mathematics in a PLC at Work Instructional Framework and Lesson-Design Evaluation Tool

  - Figure 7.2: Team Discussion Tool—Protocol for Team Analysis of the Mathematics in a PLC at Work Lesson-Design Tool

- *Mathematics Homework and Grading in a PLC at Work*

  - Figure 1.1: Team Discussion Tool—Review of Current Independent Practice Routines

  - Figure 1.2: Team Discussion Tool—Evaluating Quality of Independent Practice Assignments

  - Figure 6.4: Team Discussion Tool—Recommendations for Determining a Final Grade

After you lead the mathematics team's planning discussion using figure 2.11, you can use figure 2.12 (page 40) to clarify the SMART goal and the supporting action steps.

SMART goals should serve as a living document. What we mean by this is that team members constantly refer to and monitor the goals and action steps by analyzing evidence of student learning, with both mid-unit and end-of-unit assessment evidence. Unfortunately, SMART goals can become another checklist item if the culture of continuous reflection does not become part of the team actions. Intentionality is key.

How often does your team (or teams) currently review its SMART goals and action steps? Ideally, it should review the action steps frequently to guarantee two things.

1. Each mathematics team is completing the agreed-on plan.

2. Each mathematics team is looking at evidence of student learning (student work and instructional data) to monitor team progress on meeting the SMART goal. (See part 2 on page 59 of this book.)

Consider the grade 7 mathematics team SMART goal example in figure 2.13 (page 41). As you read through the SMART goal, identify team actions that support looking at evidence of student thinking.

When teams create SMART goals, it is important you provide feedback regarding the clarity of the goal and action steps they devise. When providing feedback, consider the following.

- How SMART is the goal? Is it written as a from/to statement?

- Does the team include short- and long-term goals? Does the team clearly articulate the evidence of success, and how will the team monitor its progress?

- What high-quality team actions (listed on page 39) is the team developing?

- Is there a focus on improved instruction and attention to the quality of the assessment and processes teams and students undertake as a result of assessment feedback?

- Do the action steps align with the site-level and district-level goals?

- Does the team clearly name a team member responsible for each action step?

Six High-Quality Team Actions	Evidence to Support High Engagement in This Action	SMART Goal Action Steps We Need to Fully Engage the Team
1. Develop high-quality common assessments for the agreed-on essential learning standards.		
2. Use common assessments for formative student learning and intervention.		
3. Develop high-quality and essential lesson-design elements.		
4. Use the lesson-design elements to provide formative feedback and encourage perseverance.		
5. Develop and use high-quality common homework assignments for formative student learning and perseverance.		
6. Develop high-quality common grading components and use formative grading routines.		

Figure 2.11: Team discussion tool—Leading team reflections on goal-setting actions.

*Visit **go.SolutionTree.com/MathematicsatWork** for a free reproducible version of this figure.*

SMART Goal			School year: _____
Department:		**Team:**	
Team leader:			
Team members:			

Identify a student achievement SMART goal (strategic and specific, measurable, attainable, results oriented, and time bound):

Action Steps and Products What steps or activities will you initiate to achieve your goals? What products will you create?	**Team Members** Who is responsible for initiating or sustaining the action step or product?	**Time Frame** What is a realistic time frame for each step or product?	**Results and Evaluation** How will you assess your progress? What evidence will you use to show you are making progress?

Source: Adapted from DuFour et al., 2006, 2010.

Figure 2.12: SMART goal template.

*Visit **go.SolutionTree.com/MathematicsatWork** for a free reproducible version of this figure.*

Additionally, when your teams continuously monitor their SMART goals (by reporting directly to you on their progress every unit), members can revise action steps if they do not positively affect student learning, or if additional challenges occur when monitoring student learning. Collective agreements and SMART goals are a great starting point for clarity of team actions toward your vision for instruction, assessment, grading and

homework, and intervention. Teams will also need to create *products* to collectively respond to the four critical PLC questions.

Once you clearly articulate the mission, vision, values, and goals, as the foundation for effective collaboration, you need to also define the work of your team that aligns with the mission and vision.

LEADER *Reflection*

What advice would you give to the grade 7 mathematics team in the example in figure 2.13 on its SMART goals and action steps? What feedback would you provide to improve the teams' plan?

_____

_____

_____

_____

_____

_____

_____

_____

SMART Goal			School year: 2020–2021
Department: Mathematics		**Team:** Grade 7 mathematics team	
Team leader: Melissa Smith			
Team members: Andre Wise, Emma Moss			
Identify a student achievement SMART goal (strategic and specific, measurable, attainable, results oriented, and time bound): First semester—We will increase the ABC passing rate from 65 percent (December 2019) to 70 percent (December 2020). Second semester—We will increase the ABC passing rate from 62 percent (May 2020) to 68 percent (May 2021).			
Action Steps and Products What steps or activities will you initiate to achieve your goals? What products will you create?	**Team Members** Who is responsible for initiating or sustaining the action step or product?	**Time Frame** What is a realistic time frame for each step or product?	**Results and Evaluation** How will you assess your progress? What evidence will you use to show you are making progress?
Focus on differentiation: Identify strategies for using formative assessment processes to create an in-class intervention (core instruction). Implement new strategies during common-lesson and team-design time. We will co-plan one lesson a unit and observe each other and provide feedback to our peers.	Melissa: Unit 1 Andre: Unit 2 Emma: Unit 3 Melissa: Unit 4 Andre: Unit 5 Emma: Unit 6	Complete first three units before December 1. Complete last three units before May 1.	The completed lessons and observation will help assess progress. New strategies for differentiation will increase student achievement on unit assessments.
For intervention, identify proficiency of each essential learning standard and identify student-by-student and skill-by-skill needed interventions. Complete a data dialogue at the end of each unit to identify specific interventions to support continued learning.	Team leader: Collect student work and lead a data dialogue at the end of every unit.	Complete the units using the following schedule. • Unit 1: September • Unit 2: November • Unit 3: December • Unit 4: February • Unit 5: April • Unit 6: May	Completed data dialogue and intervention plan for upcoming units are assessment check and evidence of progress.
Re-evaluate assessments and their alignment with mathematics standards and create student-friendly targets and common scoring rubrics.	Melissa: Unit 1 Andre: Unit 2 Emma: Unit 3 Melissa: Unit 4 Andre: Unit 5 Emma: Unit 6	Complete the units using the following schedule. • Unit 1: Before school starts • Unit 2: September • Unit 3: November • Unit 4: December • Unit 5: February • Unit 6: March	Revised assessment with student-friendly targets and completed scoring rubrics are assessment check and evidence of progress.

Figure 2.13: Sample grade 7 collaborative team SMART goal template.

TEAM RECOMMENDATIONS

Focus on Evidence of Success

With your team, consider the following recommendations.

- Create both short-term and long-term goals for teams to continuously reflect, refine, revise, and celebrate.

- Set time lines and develop a shared understanding of how the action steps will support meeting the goals.

- Team members work collectively to ensure that the team's goals and action steps are aligned to the mission and vision of the school or district.

Clarity and Reflection On What Is Tight and Loose

As your teams establish and understand a common vision for their work, develop collective commitments that focus on collaboration toward that vision, and identify how they will track evidence of success using goals, you begin to closely examine if your teams are, in fact, doing the right work. That is, are your mathematics teams focused on the right wisdom and work that will significantly impact student learning?

Remember, your first leadership and coaching action is to develop PLC structures for effective teacher team engagement, transparency, and action. Your role as a leader is to be crystal clear on the collaborative learning structures your teams need for effective engagement, transparency, and action.

If mathematics teams have an action step related to assessment, as a team leader, how do you know the quality of the assessments the team uses? As a coach, what is the professional learning you provide to develop the teachers' assessment literacy? If there is an action step in the plan related to instruction, what content knowledge or pedagogical knowledge should your teams develop before they can take that next step in their lesson design? Or, are there mathematical content strategies your teams need to explore? Are there grading inequities your teams need to address?

The answers to these questions require different solution pathways and artifacts that your teams of mathematics teachers work together to navigate. By using a tight-loose leadership model, you can provide clear expectations for the collaboration in which teams will engage as they examine solutions and artifacts.

Tight-loose leadership is a combination of centralized leadership (tight—we will all do this action) combined with participative decision making and freedom to make individual decisions (loose—freedom for how to do the tight action). In a PLC, team members need to understand what is tight (expectations for all) and what is loose (the solution pathway) for improvement (DuFour et al., 2016; Kanold, 2011).

As a leader, you set the clear expectations for what is tight on the teams you support and what is loose. Using figure 2.14, reflect on the current expectations for the collaborative teams you lead or support and define actions that are tight and loose in relation to the topics in the first column.

	What Is Tight?	What Is Loose?
Common unit assessments		
Instructional strategies		
Mathematical tasks chosen		
Lesson-design criteria		
Common homework		
Common grading		
Interventions during core instruction		
Interventions to support core instruction		

Figure 2.14: Team discussion tool—Tight-loose aspects of team actions.

*Visit **go.SolutionTree.com/MathematicsatWork** for a free reproducible version of this figure.*

After you help team members complete the figure 2.14 reflection, compare your responses to the sample team response provided in figure 2.15. What are the similarities or differences with your personal team reflections? Does your team agree and understand the what and the why for what is loose and what is tight?

When analyzing the examples in figure 2.15, look at the team agreement on lesson design for an example. Notice the tight aspect of lesson-design criteria: address the common lesson components for instruction. Teams together explore the effective lesson-design components in the Mathematics in a PLC at Work lesson-design tool (available to download from **go.SolutionTree.com /MathematicsatWork**), discuss the common lesson elements, and come to agreement on a possible instructional activity or technology that best supports the learning target for the lesson. (See also Kanold, Kanold-McIntyre, et al., 2018.)

However, the loose aspect of lesson design also takes into account the personality and style of each and every teacher. So, while the lesson elements may be the same, the way educators teach the lesson may be different. The key to understanding the tight-loose leadership of lesson design is to ensure that the same essential learning criteria are the focus of the lesson and the individual team members equally honor the expectations for rigor.

	What Is Tight?	**What Is Loose?**
Common unit assessments	• Give students' agreed-on common mid-unit and unit assessments. • Administer each common assessment in an agreed-on way in each class. • Discuss student results on common assessments and look at actual student work as examples.	Create and administer any common mid-unit assessments in addition to the common unit assessments.
Instructional strategies	• Create lessons and instructional resources that support the teams' instructional focus for student perseverance.	Choose instructional strategies to support individual student needs.
Mathematical tasks chosen	• Collectively choose performance tasks to use in each unit and reach agreement on the balance of the cognitive demand of those daily tasks.	Choose additional tasks as needed to meet individual students' needs.
Lesson-design criteria	• Address the six common lesson components for instruction in the Mathematics in a PLC at Work lesson-design tool (available at **go.SolutionTree.com/MathematicsatWork**).	Design how each lesson is taught using those components.
Common homework	• Collectively choose the best independent practice that is massed and spaced.	Make adjustments to assignments to meet individual student needs and special circumstances.
Common grading	• Grade common mid-unit and unit assessments in an agreed-on way (for example, determine scoring points for each question and how you will allocate those points for partial credit). • Discuss student results by standard on the common unit assessments and look at actual student work as examples.	Nothing about the scoring of the common assessments should be loose.
Interventions during core instruction	• Using student work from common assessments and daily tasks, together analyze current Tier 1 strategies to improve upon in your lesson designs.	• Implement classroom interventions. • Implement classroom extensions.
Interventions to support core instruction	• Using student work from common assessments and tasks, establish a Tier 2 collective and targeted response to student learning by essential mathematics standards for each unit.	Although the teacher team works together to ensure every child receives the same required intervention and support, it is possible that individual students receive additional support as needed.

Figure 2.15: Sample tight-loose aspects of team actions.

LEADER *Reflection*

What aspects and actions for teaching and learning mathematics do you view as tight versus loose?

_____   _____

_____   _____

_____   _____

_____   _____

_____   _____

Once your team understands what is tight, you help team members move to establish team agreements about the right work and how to accomplish it. Figure 2.16 provides a protocol for reaching agreement on the work of a mathematics collaborative team.

Use the team-building worksheet (figure 2.16) at the beginning of each school year in conjunction with creating your collective team commitments, setting SMART goals, and developing action steps. If new team members join the team mid-year, expect the team leader to share the team-building worksheet agreements with the new members to (1) eliminate excuses for not knowing the expectations and (2) create an open line of communication between veteran team members and new team members. Be willing to revise as you need. Review these artifacts with each team at least three times per year, if not more often. The protocol provided in figure 2.16 supports mathematics team understanding of the right work to do, and provides clarity on:

- *What* artifacts the team needs to develop or revise

- *Who* is responsible for what artifacts

- *When* the team needs to develop the artifacts

- *Where* the team will engage in the work

- *Why* the work of the team is critical to the SMART goal and vision for assessment and instruction

- *How* the team will effectively work together

After your team completes the team-building worksheet, team members should be able to clearly articulate any additional action steps and prioritize team actions they may need to take to meet the team goals.

Figure 2.17 (page 46) provides a list of sample artifacts or products your mathematics teams may include as part of their work and action steps in their SMART goal plan. With your team, review each of the artifacts and discuss its purpose along with what critical questions creating such an artifact would answer.

As you review the possible team artifacts in figure 2.17, discuss actions your mathematics teams need to include in their action steps for this current year, and the actions they need to build into plans in the next two to three years.

TEAM RECOMMENDATIONS

Develop Protocols for Clarity and Reflection on the Right Work

With your team, consider the following recommendations.

- Leaders and team members understand the tight and loose expectations of their teamwork.

- Leaders provide feedback to the team on the team's action steps and teamwork.

- Team leaders create intentional opportunities (protocols) for teams to communicate expectations of specific teamwork and the who, what, when, where, and why of the work.

Team-Building Protocol Worksheet

School Year: _____   Team Name: _____

Team Leader: _____

Team Members: _____

_____

_____

_____

Part 1: Organization Details

Address the following about organization details.

- **Meeting frequency:** When and where will we meet for our team meetings each week or month?
- **Extra meetings:** When and where will we meet for additional meetings and how often?
- **Vision for student achievement:** What are our student-achievement goals for the current school year based on our vision for instruction and assessment? (Complete the SMART goal template from figure 2.12, page 40.)
- **Team strengths:** List the strengths each team member brings to our team.
- **Team collective agreements:** What are our team's collective commitments? What are our team norms to support effective collaboration?
- **Feedback on our collaborative team actions:** Discuss our expectations for feedback and appreciation. How do we envision the team functioning throughout the year?
- **Conflict:** State a procedure we will use when discussing conflicts that will naturally occur throughout the year.

Part 2: Essential High-Quality Protocols and Effective Monitoring

Address the following about high-quality protocols and effective monitoring.

- **Common guaranteed and viable curriculum:** Describe our team's work around the essential learning standards and the content and the processes that we expect all students to know and be able to do. Name the person or people on our team responsible for ensuring all team members are implementing the curriculum with fidelity.
- **Common assessments and scoring:** Describe our team's work around the responsibility for and implementation of a high-quality system for our common assessments. Name the person or persons on our team responsible for test implementation throughout the year.
- **Common independent practice (homework) assignments:** Describe our team's work around the responsibility for and implementation of a high-quality system for common independent practice. Name the person or people on our team responsible for homework implementation throughout the year.
- **Common higher-level-cognitive-demand tasks:** Describe our team's work around the responsibility for and implementation of a high-quality system for using common in-class high-cognitive-demand tasks for each unit. Name the person or persons on our team responsible for task implementation throughout the year.
- **Common feedback (scoring response) and unit reflection:** Describe our team's work around the responsibility for and implementation of a high-quality system for our common response to student learning at the end of each unit. Name the person or persons on our team responsible for our end-of-unit response.

Figure 2.16: Team-building protocol worksheet.

*Visit **go.SolutionTree.com/MathematicsatWork** for a free reproducible version of this figure.*

You and your collaborative team are now ready to begin the work! Armed with a strong set of collective agreements to guide team actions, a SMART goal, and an action plan as a GPS for the team, you will now explore the final leadership strategy: mutual accountability to the professional work of the team.

Artifact	Purpose
Common course or grade-level syllabus for each mathematics unit	
Unit plan (including identification of the essential learning standards and agreement on the learning targets for the unit)	
Common unit assessments with common scoring rubrics for those assessments	
Common mid-unit assessments with scoring rubrics for those assessments	
Common unit independent practice assignments (homework)	
Common review materials or instructional resources for the unit assessments	
Instructional strategies for each standard in the mathematics unit of study	
Shared and common use of technology tools for each unit	
Lab or project materials used for each unit	
Common grading expectations for makeup work	
Common feedback protocols for student reflection and improvement of grades	
Common lesson-plan designs and discussions	
Common analysis of student work or data using a specific protocol or process for the purpose of interventions	

Figure 2.17: Possible team artifacts and products to produce.

*Visit **go.SolutionTree.com/MathematicsatWork** for a free reproducible version of this figure.*

Mutual Accountability to the Professional Work of the Mathematics Team

Products that mathematics collaborative teams create and the processes that your teams engage in are critical to building a PLC culture. However, at the heart of the PLC life are the relationships that are built and the conversations that occur during your team meetings. The professional responsibilities of team members, team leaders, and other leaders who support a team are to collectively move beyond the *sharing* of your work, and into the responsibilities of and accountability for improving mathematics teaching and learning, together. Consider the kindergarten team scenario in figure 2.18. As you read through the scenario, identify the issues that exist with this team's current level of collaboration. Has this team become *interdependent* with one another?

You may have noticed some of the following strengths of the kindergarten team in figure 2.18: designated team time to meet, agendas, and SMART goals and norms.

You may have also taken note of some of the challenges present on the team or with the team leader's actions.

- Teachers are unwilling to implement the new instructional resources.

- Teachers are not willing to try something new or take risks.

- Members don't follow the agendas.

- No one monitors norms, or no norms are established to address conflict resolution.

- Team members do not hold each other accountable for *implementing* the agreed-on essential standards.

- Unintentional inequalities continue to prevail.

LEADER *Reflection*

If you were the coach or site-level leader for the kindergarten team in figure 2.18, what feedback would you provide to move them forward? If you were their team leader, what next steps would you suggest to move the team forward?

_____

_____

_____

_____

_____

_____

_____

_____

_____

Kindergarten Mathematics Team Scenario

The kindergarten team consists of five kindergarten teachers and a special education teacher. They had been meeting for four weeks for one sixty-minute period each week. The team created its SMART goals and established norms, yet the team leader was struggling to get the teachers to work as a team or come to agreement over the first month. The team completes few of the action steps it has planned.

After the first four weeks of ineffective meetings, the team leader talks to the teachers separately to try to understand what is causing the tension. The problem becomes immediately obvious: three of the team members are used to the old instructional resource and two newer team members attended the new curriculum workshop over the summer and are excited to use the new instructional resources for mathematics. The special education teacher, who is friends with one of the more traditional teachers, does not want to ruin their friendship, even though she is coteaching with one of the newer team members and sees a huge difference in student understanding using the new instructional resources.

During team meetings, the new members share their experiences using the new instructional resources. The other team members disregard what the new teachers are sharing and continue to utilize the materials from years past.

The team leader does create an agenda and distribute it to the team members prior to the team meetings, but since heated discussions are derailing the meetings, the leader never completes or shares minutes with the instructional coach or principal. Everyone knows that as soon as the meeting is over, the teachers will retreat to their classroom and choose to use a resource that they prefer over what the team is trying to agree on.

Figure 2.18: Kindergarten mathematics team scenario.

What do you notice about the current level of responsibility of each team member on the kindergarten team? Who is responsible for holding the team members accountable for their actions (or inactions)? The kindergarten team leader struggles to create a structure to support accountability for improving the mathematics teaching and learning of each and every student.

If the collective agreements for the kindergarten team had included a norm for reaching consensus, emotions might not have taken over the meeting. If the group had identified a team member to monitor the teams' behaviors, members may have been able to overcome their differences.

When you lead or support mathematics teams, it is your responsibility to ensure shared responsibility of team actions. Each and every team member (and the leader) must understand:

1. What work needs to be done

2. How the work will be done

3. When the work needs to be done

4. Where the work will have impact

5. Why the work is important

You can use the following structures to develop shared understanding of a mathematics team's work. The structures also support building shared accountability and transparency measures, specifically:

- Agendas and minutes for each meeting

- A time line for the cycle of inquiry and action research

- Expectations for how the team will share its work among team members

- Expectations for the quality of the work and how the team monitors it

The following descriptions provide more detail about each of these shared accountability and transparency measures for your teacher teams in mathematics.

Agendas and Minutes for Each Meeting

Ever feel lost at a meeting that does not seem to have a purpose or an end time? Ever been to a meeting with no clear agenda? You might have felt frustrated due to a lack of team focus that resulted in no decisions, agreements, or conclusions being reached by the end of the meeting.

These nonexamples of productive meetings emphasize the importance of a plan. Most mathematics teams have set times each week to meet. If you are the team leader, it is your responsibility to ensure that the time is well spent; it is the professional responsibility of each team member to provide support and suggestions for the team agenda on how to effectively use the time. If you are the coach or site-level leader, it is your responsibility to give teams feedback on the planned agenda—not to create the agendas for the teams.

In figure 2.18 on page 47, the kindergarten team has an agenda, created by the team leader. One reason the team struggles to follow the agenda is that the leader does not consider or request team member input.

Make sure, however, that the team's work focuses on meeting the SMART goal and vision for your work. If you notice the team input is outside the scope of the team's action steps, ask the team for clarification on how the time it spends meeting as a team will support the SMART goals.

Figure 2.19 is a sample agenda from a high school geometry team. As you read it, identify the components of the agenda that create transparency.

The essential components for team transparency and accountability include:

- Informing all members when the meeting will occur

- Identifying who is present at the team meeting (and who is not present)

- Providing focus on how the team is addressing the four critical questions

- Addressing time constraints by including how much time will be spent on each topic

- Referring to the current SMART goal to promote continual reflection and alignment of team actions

- Listing a clearly defined outcome and a record of what the team is accomplishing

- Clarifying next steps for each team member and for the team

Mathematics Department Agenda and Minutes		
Date and time: January 9, 2017 from 7:30 a.m.–8:25 a.m.	**Team:** Geometry team	**Team leader:** Ms. Berg
Team members in attendance: Mr. Shinkle, Ms. Mays, and Ms. Cunningham	**Other participants:** District mathematics coach	
Current SMART goal: By the end of the 2016–2017 school year, the percentage of students proficient on the state assessment in mathematics will increase from 75 percent to 82 percent.		
What Critical Question or Questions Will We Focus on at This Meeting?		
☒ 1. What do we want all students to know and be able to do? ☒ 2. How will we know if they learn it?	☒ 3. How will we respond when some students do not learn? ☒ 4. How will we extend the learning for students who are already proficient?	
Actions and Agenda Items	**Minutes of the Meeting**	

1. **Semester exam discussion (ten minutes):**
  • Review the exam blueprint.
  • Discuss what instructional resources we are going to use for the warm-ups for the final six weeks.
 	
2. **Unit assessment (fifteen minutes):** Share feedback on the upcoming unit 4 assessment. (Ms. Mays will put it in the shared folder.)
 	
3. **Mid-unit data dialogue (twenty minutes):**
  • Bring in-class sets of student work for the unit 3 quiz. (Mr. Shinkle will create a Google Form for the analysis.)
  • Determine essential standards for advisory tutoring.
 	
4. **Closure (five minutes):** Identify action steps for the next meeting and what needs to be completed this week.
	The team discussed the semester exam final blueprint and made sure that the current pacing would provide additional time and support for the concepts that students are still struggling with. They also made sure that they were including multiple select-response items on warm-ups and independent practice to better prepare the students for the new item types on the semester exams.
	
	The team reviewed the draft of the unit 4 assessment, and Ms. Berg is going to make the edits and place them in the shared folder.
	
	The team reviewed the data analysis from the unit 3 quizzes and planned the re-engagement activity for advisory. Ms. Cunningham is going to work with the tutors during advisory to make sure that the tutors can support the student questions.	

Next Steps for the Team	**Agenda Items for Next Time**	
For next time: Create a reassessment for the unit 3 targets that we are reteaching during advisory. Continue to discuss the instructional resources we are going to use for warm-ups.	1. Create reassessments for unit 3 targets. 2. Discuss instructional resources for the warm-ups. 3. Look at the unit 4 assessment data and plan unit 4 interventions.	

Figure 2.19: Sample geometry team meeting agenda.

*Visit **go.SolutionTree.com/MathematicsatWork** for a free reproducible version of this figure.*

Use the leader reflection to take a moment to reflect on your structures for team agendas and minutes.

LEADER *Reflection*

What structures do you currently have in place to create mutual accountability and feedback to the team meeting agendas and minutes?

_____

_____

_____

_____

_____

_____

_____

_____

You can use technology to share agendas and minutes with all team members via Google Docs (https://docs .google.com), Dropbox (www.dropbox.com), OneDrive (https://onedrive.live.com), or other shared file sites. This allows your teachers to access the document if they forget an agreement or are absent from the team meeting. When you are transparent with minutes kept for each mathematics team meeting, there are no excuses for members not adhering to team agreements and commitments. To increase team meetings' effectiveness, you will then need to establish a structure for shared agendas and minutes.

Being transparent also provides you with a vehicle for opening lines of communication between team members and all mathematics leaders to support the collaborative process. If you are a site-level leader or coach who does not attend every meeting, ask your teams to submit minutes and then review the minutes and provide feedback. If you are a team leader and are not able to be at a meeting, be sure to assign someone on the team to collect the evidence of team learning. Minutes are an opportunity for you to provide feedback to your teams to promote and further the collaborative process.

In addition to creating minutes and agendas for your mathematics teams, you also need to create a focus on how a team collectively responds to student learning using the four critical questions of a PLC and the six team actions we explore in *Mathematics Assessment and Intervention in a PLC at Work*, *Mathematics Instruction and Tasks in a PLC at Work*, and *Mathematics Homework and Grading in a PLC at Work*.

A Time Line for the Cycle of Inquiry and Action Research

As soon as one mathematics unit ends, another begins. It is a continual cycle. As a leader, you understand that inquiry and action research can happen before, during, and after a unit of mathematics instruction. How do you make sure mathematics teams are using the six team actions from the Mathematics in a PLC at Work framework to answer the four critical questions of a PLC in *every* unit? How much time do team members spend creating common unit assessments? Choosing tasks and homework? Looking at student work? Planning interventions or enrichments? What you may find is that teams spend more time on PLC critical questions one and two and less time on critical questions three and four, or vice versa.

The key team action at the end of a unit is to generate and analyze data to make informed decisions about next steps for your team and address the time line for the cycle of inquiry and action research to follow. Read the sample time line for a three-week unit of instruction in figure 2.20. As you read through the time line, consider the team actions and supporting teacher leader actions that occur during each collaborative team meeting.

Notice the intentional planning of the team actions to ensure there is a collective response to student learning. The team consistently addresses the critical questions during the cycle of inquiry, and there is a clear connection between the four critical questions and the six team actions in the *Every Student Can Learn Mathematics* series.

Unit Time Frame	Team Actions	Team Leader Actions	Coach and Site-Level Leader Actions
Before the unit	Identify the essential standards. Create common unit assessments with agreement on scoring rubrics and feedback. Determine common independent practice. Discuss Tier 1 instruction plans.	Schedule time to make sure each team member has a collective understanding of the essential standards as well as the assessment evidence and instructional strategies that support student learning.	Know when your teams are starting a unit of instruction and provide time, support, and feedback for building shared knowledge of the unit expectations. Facilitate meetings as you need to.
Week 1 meeting	Review the current scope and sequence (order of the standards) in relationship to student learning—how is it going? Analyze data and evidence of learning from a during-the-unit assessment to inform Tier 1 and Tier 2 instruction. Identify any common lesson expectations and tasks.	Discuss current student learning for each essential learning standard, and plan for reengagement and enrichment strategies.	Provide feedback to teams as they analyze student learning from common assessments and tasks during the unit. Support with identifying effective instructional practices or quality tasks may be necessary.
Week 2 meeting	Review the current scope and sequence in relationship to student learning—how is it going? Analyze data and evidence of learning from a during-the-unit assessment to inform Tier 1 and Tier 2 instruction. Create common unit assessments for the next unit of study and determine the common independent practice. Clarify assessment procedures for current end-of-unit assessments.	Discuss current student learning on essential learning standards, and plan for reengagement and enrichment. Review upcoming unit assessment, and come to agreement on proficiency, assessment evidence, and common artifacts.	Provide support for Tier 2 interventions. Provide feedback to teams when they are monitoring student learning prior to the common assessment.
Week 3 meeting	Analyze data from the end-of-unit common assessment and plan for intervention and enrichment. Review new unit scope and sequence. Agree on the scoring rubrics and feedback for common assessments in the new unit. Discuss Tier 1 instructional practices.	Analyze student learning to make instructional decisions, and provide feedback to students, which requires action.	Know when your teams are ending a unit of instruction and provide time, support, and feedback on instructional plans to support each and every learner.

Figure 2.20: Sample cycle of inquiry for a three-week unit of instruction.

How do you currently create intentional opportunities to engage your team in consistently answering the four critical questions of a PLC?

_____

_____

_____

_____

_____

_____

_____

_____

You can use your minutes from the team meetings for each unit of instruction to reflect on how much time the team is engaged with each Mathematics in a PLC at Work team action. You can also utilize the checkboxes approach as on the sample agenda (figure 2.19, page 49). By checking a box next to the critical question that is the focus each time the team meets, you can use data at the end of a quarter or grading period and tally up

how much time the team has spent addressing each area to inform future work with the team.

However, a word of caution when you use checklists: The agenda example in figure 2.19 includes a checklist for the four critical questions. You do not want these boxes to be just another item for the team members to check off as *done* each week. You will need to clarify which team actions support answering each critical question. When you lead your mathematics team to reflect on which questions they are collectively responding to, you can then use their reflection to provide feedback. You can also note which questions the team is consistently addressing to ensure your teams are balancing the work and moving to effective collaboration.

Expectations for How the Team Will Share Its Work Among Team Members

As a leader, do you know the reflect, refine, and act professional work of each mathematics team? Do you know each team's story? For example, imagine a reporter contacts you and wants to understand the work your teams have done to increase the percentage of students enrolled in your grade 8 algebra course (or any advanced course at your school) and how the demographics of students enrolled in this course are the same or different from the demographics of the district as a whole.

Personal Story **MONA TONCHEFF**

Each year, our collaborative teams set two primary SMART goals. One goal was to increase the percentage of students passing the state assessment, and the second goal was to increase the weekly percentage of students passing each mathematics course with a C or better. During the first few years, I noticed that every year, we had to search to find the data. It was not readily available, and when writing a specific SMART goal, the teams needed to include the current reality (the *from* point) and where the teams want to be at the end of the goal (the *to* point). When teams couldn't locate their data, the goals were not as specific as they should be.

To add to the challenge, our collaborative mathematics teams were not looking at the trends in student performance. As the leader, I had to create a structure of where to locate the data and then to share these data. I needed to specify the expectation my teams would monitor and then respond to the trend data throughout the year. For mutual accountability, I had to model how to explain our year-to-year story, and then show the collaborative teams how to use the trend data to inform their individual team actions.

The reporter would like to know how many students were in the course five years ago compared to how many there are today, and what specific actions you and your team took to increase not only the number of students in the course but also increase student achievement and improve the demographic alignment in the course to that of the overall district. The high school the students attend after the K–8 school has also observed an increase in students taking advanced mathematics classes, and the community likes to hear a good success story.

The key to an articulate and positive response to this reporter is to ensure your mathematics teams are setting SMART goals for these types of data, working together to monitor evidence of student participation in these courses as well as their learning, and then creating the necessary artifacts and processes to meet each SMART goal. Thus, your teams should simultaneously be archiving the work and responding to the observed results.

Creating transparency without shame in using the data is important. When you are transparent with the work and history of the team, you are building shared knowledge of past actions to identify new team actions to take.

If new team members join a team, they should have ready access to a shared data learning space in order to know student results from the past. Archiving work so it is easily shared also allows your teams to recognize trends in student achievement.

However, when pulling data reports, will you help your teams compare student performance from year to year? Does the data representation display trends over time? Most site-level and district-level data are easily configured for one specific range of time. If you want mathematics teams to monitor and respond to their team progress from year to year and unit to unit on their agreed-on measure of success, you need to create a structure for gathering historical data. Figure 2.21 is an example of two types of measures for success collected over time.

Not only can you collect data for each year; you can also collect data over time to see the difference from year to year and unit to unit. Teams can also collect data from unit assessments over time to monitor team growth with specific units and standards. When reflecting on your current system for data collection, use the following leader reflection to consider the additional structures you would need in place to fully picture your team story.

Grade Level or Course and Year	Grade Distribution					Unit 1 Assessment Passing Rate			
	A	B	C	D	F	Pass		Fail	
						Number	Percentage	Number	Percentage
District									
School A									
School B									
School C									

Figure 2.21: Sample data trend report.

What data do you use to help your mathematics teams analyze student learning at the end of every mathematics unit and at the end of each year?

When analyzing these data, what structures could you create to help the teachers gather data trends and efficiently look at trend data from the past three to five years?

_____

_____

_____

_____

_____

_____

_____

_____

_____

_____

_____

_____

_____

_____

_____

The final structure for building shared accountability and transparency is for you to monitor and provide feedback on the teams' professional work. How do you focus their efforts on high-quality mathematics work?

Expectations for the Quality of the Work and How the Team Monitors It

Another professional responsibility of a mathematics leader is to effectively monitor all team-produced artifacts and build capacity on what is considered high-quality work. Ensuring that each team member has access to the shared and agreed-on artifacts (such as mathematics homework assigned) is important, but equally important is to discuss with each the quality of the homework assigned. Does it follow a research-affirmed base?

For example, when the team creates common mathematics unit assessments, how do the team members know the assessment gathers the best evidence of student learning?

When the team designs some lessons together, what evidence of student engagement, learning, or summary will be used to evaluate the effectiveness of the lesson?

When the team discusses intervention and grading procedures and make-up routines, how do teachers know the process they use for their feedback to students aligns to research-affirmed practices?

In this *Every Student Can Learn Mathematics* series, you are provided three additional books with specific team rubrics you can use to help your team align to the six team actions described in the Mathematics in a PLC at Work framework in figure I.2 on page 4. Use those evaluation of current practice rubrics to guide your teams' conversations on the quality of their work together. Access these tools online at **go.SolutionTree .com/MathematicsatWork**.

In your role as leader, you identify the collaborative team actions needed for improvement and the support you provide to move the team forward. It is best to *focus on one aspect of quality* at a time. Effective collaboration takes time and practice. If you want the quality of the assessments, instruction, grading, homework, or intervention products and routines to improve, reflect (and eventually refine) on the quality of your teams' work one school season at a time.

Engaging your teams in meaningful reflection is a vital leadership strategy you use to improve collaborative efforts. You can use figure 2.22 to ask each team member to reflect on the three big ideas of a PLC. They should complete the reflection individually, and then bring their results to a team meeting for discussion.

Mathematics Collaborative Team Reflection

Directions: Rate your collaborative team level related to the three big ideas of a PLC.

Grade-Level Mathematics Team Self-Assessment		Rubric Score			
		Level 1 Beginning	Level 2 Practicing	Level 3 Implementing	Level 4 Embracing
Focus on Collaboration	My mathematics team moves toward a collective vision and creates norms that reflect the collective commitments to the work.				
	My mathematics team creates and uses an action plan to meet its SMART goals.				
	My mathematics team is productive and works cohesively to meet the common learning outcomes.				
Focus on Learning	My mathematics team makes sense of the essential learning standards in each unit and creates student-friendly learning targets, which students use as a reflection tool throughout a lesson and unit.				
	My mathematics team designs quality common unit assessments that meet the rigor of the standards, and creates common scoring agreements for each assessment.				
	My mathematics team analyzes data from common assessments by standard and then plans for collective student re-engagement in learning via class instruction or an intervention time.				
	My mathematics team identifies the instructional strategies that impact student learning for replication in future units or next year. We regularly discuss how to actively engage the students in peer-to-peer discourse for at least 50 percent of the mathematics lesson.				
Focus on Results	My mathematics team discusses how students earn grades and ensures students earn grades consistently from teacher to teacher based on proficiency versus effort.				
	My team creates common homework assignments for each unit provided to the students when the unit begins.				
	My mathematics team consistently identifies students who do not meet, meet, and exceed essential learning and collectively creates a plan to ensure learning standards for all students.				
	My mathematics team engages in lesson studies and observes each other to discuss how best to engage students in the learning of mathematics.				

Of the eleven criteria, identify which are your areas of strength as a teacher team of mathematics and explain why.

Of the eleven criteria, identify which are areas of needed growth for your team. Identify next steps to move your team toward level 4.

Figure 2.22: Team discussion tool—Mathematics collaborative team reflection on the three big ideas of a PLC culture.

Visit **go.SolutionTree.com/MathematicsatWork** *for a free reproducible version of this figure.*

When you and your team meet to discuss the reflection using figure 2.22 (page 55), look for areas of strength and needs to identify certain aspects of the PLC process that you should refine.

TEAM RECOMMENDATIONS

Establish Mutual Accountability

With your team of teachers of mathematics, consider the following recommendations.

- Establish a structure for consistent communication of mathematics team agendas and help the teachers learn to summarize the work of the team.

- Establish a time line of the work of the team to ensure the mathematics team is focused on answering all four critical questions of a PLC culture.

- Create a repository of team artifacts and a history of the student data performance for each mathematics team.

- Ensure there is a research-affirmed quality to the common artifacts and provide feedback with team meeting support as needed to improve that quality.

The five leadership strategies for mathematics team engagement are:

1. Create a common purpose of each mathematics team (teachers of mathematics and support personnel as well)

2. Develop collective commitments by the team that focus on collaboration

3. Identify evidence of success; specifically, focus team members on results and gather evidence over time of student mathematical thinking and learning

4. Gain clarity and reflection on the right research-affirmed work for maximizing student learning

5. Provide structures that promote mutual accountability to the professional work of the team

When you combine the leadership strategies with the personal leadership practices described in chapter 1, you can develop and honor the work of your mathematics teams simultaneously. You also create meaningful teacher professional learning opportunities through highly effective collaboration in order to subsequently impact the learning for each and every student. You have now established the foundation for ensuring equity for student mathematics learning and outcomes.

In part 1 of this book, you addressed personal leadership practices and strategies for effective collaboration. These leadership practices and strategies, when fully developed, have a magnified impact on student achievement and help close learning differentials. You start your mathematics leadership work by building shared mission (your purpose), vision (what you want to become), and values (your collective commitments), and then set goals (your outcomes) to drive the mathematics work of each collaborative team.

Leading others requires a passion for the work and the relational skills necessary to build a caring community of learners.

Part 1 supports your mathematics leadership efforts by responding to components of the four critical PLC questions in relation to the teams or team members you serve:

1. What do we want all teacher *team members* to know and be able to do?

2. How will we know if a teacher *team member* learns it?

3. How will we respond when some teacher *team members* don't learn it?

4. How will we extend the learning for teacher *team members* who are already proficient?

Part 1 also included protocols for establishing norms, SMART goals, and a common vision, and it addressed expectations for the work of all mathematics team members to know and be able to do.

Now that you have developed effective leadership *practices* for working with teams and *strategies* to build team effectiveness, *how* do you work with teams to meaningfully address the four critical questions of a PLC culture? Where do you start?

You will need to find an entry point as you engage teams through the routines and processes in chapter 2 (page 25) and prioritize your mathematics team's actions to create a progression of team learning. Part 2 of this book examines the protocols you can use to support your various entry points (think differentiated learning for each grade-level or course-based mathematics team) for coaching and navigating each team into continuous action research.

To create equity and access to meaningful mathematics, the protocols provided add structure to help you engage each and every team member to collectively share the responsibilities of the team and improve the student learning of mathematics in your schools.

Part 2 will also address high-quality protocols for engaging your team members in meaningful discussions on how to establish a culture of transparency and focus on learning for mathematics assessment and instruction.

Coaching Action 2:
Use Common Assessments and Lesson-Design Elements for Teacher Team Reflection, Data Analysis, and Subsequent Action

We do not learn from experience . . . we learn from reflecting on experience.

—*John Dewey*

As a leader in a PLC culture, you are expected to lead a culture of continual reflection, refinement, and action based on your professional work done in collaboration with your colleagues. The culture looks like the following (DuFour et al., 2006, 2010, 2016).

1. Educators work collaboratively.

2. Educators work in recurring cycles. (You decide how often that cycle should be.)

3. Work focuses on collective inquiry and action research to achieve better student results. (You reflect on evidence of learning in your classrooms.)

4. This work serves job-embedded adult learning. (You refine and act on the evidence.)

Thus, collaboration is a *continuous*, *job-embedded*, and *ongoing learning culture* and a process you lead for your mathematics teacher teams. Your collaborative efforts are for the purpose of developing the knowledge of each member of your mathematics teams, at each moment during the school year in which you plan to use that knowledge for the units of study you are about to teach.

This is complex work; it expects all mathematics teachers to embrace a culture of transparency. It requires teachers to willingly share evidence of learning from unit assessment data, to act on that learning, and to publicly teach in front of one another to refine their own understanding of how students learn mathematics.

This is why mathematics leading and coaching via effective collaboration are such complex yet rewarding tasks. The results in both adult knowledge and student knowledge development are unparalleled to any other leadership action you can pursue.

How do the engagement and collaboration of your mathematics team or the teams you serve magnify the impact of student achievement? How do the teachers on your team use evidence of student learning to improve instruction or assessments? How does the team create intentional opportunities for continual reflection and refinement? The answer to these questions is the purpose of part 2 of this book.

Recall the four critical questions of a PLC in relation to the teams or team members you serve (DuFour et al., 2016).

1. What do we want all *team members* to know and be able to do?

2. How will we know if *team members* know it?

3. How will we respond when some *team members* don't learn it?

4. How will we extend the learning for *team members* who are already proficient?

The tools and protocols in part 2 help you evaluate your mathematics team's effectiveness and provide specific steps for you to use to respond to team learning. You will explore structures all mathematics leaders use to engage teacher teams in continuous job-embedded learning through action research and inspection of student learning.

Part 2 establishes *processes* for teachers and team members as they seek to analyze evidence of student learning—within the cycle of inquiry that teams participate in on a unit-by-unit basis. To establish a culture of continuous learning, you have to support a balance on the time spent on the critical aspects of inquiry.

Use the leader reflection to consider the following questions about your current PLC culture and the time your mathematics teams spend on specific team actions.

Part 2 of this book highlights protocols your team can use to collaboratively analyze and respond to evidence of student learning with assessment and meaningful and relevant instruction. The three chapters of part 2 focus on processes for:

- Leading a culture of reflection, refinement, and action with your mathematics teams

- Leading a culture of transparency and learning with mathematics *assessments*

- Leading a culture of transparency and learning with mathematics *instruction*

This part also includes leader and team discussion and reflection tools designed to support your work as described in part 2.

- Leadership Reflection on PLC Structures and Protocols (figure 3.2, page 63)

- Leadership Rubric for Utilizing Protocols for Effective and Focused Collaboration (figure 3.3, page 64)

- Evidence of Effectiveness Protocol (figure 3.6, page 69)

- Lesson-Study Protocol (figure 5.4, pages 94–95)

LEADER *Reflection*

How do you effectively create and nurture a culture of change, growth, reflection, and improvement?

In what percentage of your team meetings do members reflect on the impact of the team's actions on student learning?

What percentage of time do you spend refining your team's current work based on student learning?

What percentage of your time do you spend on new actions based on evidence of student learning (or not)?

_____

_____

_____

_____

_____

_____

_____

_____

CHAPTER 3

How to Lead a Culture of Reflection, Refinement, and Action With Your Mathematics Teams

The essence of intelligence would seem to be in knowing when to think and act quickly, and knowing when to think and act slowly.

—Robert Steinberg

Armed with SMART goals, collective commitments, and action steps, what is your next leadership move for mathematics in your school? How will you measure the impact of teacher team actions on student mathematics learning? As teams cocreate unit assessments, how do you engage those teams in effective reflection and improvement? As teams co-plan a lesson, what structures can you create and use for effective reflection and improvement? To address these questions, first examine the progression of collaboration and how to effectively support continuous reflection for mathematics team growth.

Effective Collaboration

Consider the scenarios in figure 3.1 (page 62), and reflect on the similarities and differences in the teams' actions before reading any further.

The team scenarios in figure 3.1 are typical of how a teacher team might respond to analyzing student learning. The grade 5 team's general response (or lack of response) is not tied to academic learning. The newer teachers are struggling with classroom management, and

when they review their student results, they are quick to blame the student disengagement in the lessons instead of reflecting on their team actions with instruction.

However, on the grade 7 team, each member looks at what students are or are not learning in order to analyze the results. The team's action focuses on learning. Through deep analysis and reflection on student work, team members are able to learn from each other and grow as professionals through the process.

Use the leader reflection on page 62 to first reflect on your next leadership steps if you were the leader of the grade 5 mathematics team and the grade 7 mathematics team described in figure 3.1.

Then use figure 3.2 (page 63) to reflect on the current reality regarding the PLC structures appropriate to your leadership role that have been implemented for each mathematics team.

Next, use the team discussion tool in figure 3.3 (page 64), the leadership rubric for utilizing protocols for effective and focused collaboration, to rate how well your teams implement and use each team protocol.

Grade 5 Team	Grade 7 Team
The four teachers in the grade 5 mathematics team have been engaged in the PLC process for four years. This school year, three of the team members are moving to new teams, and the team leader has three new team members. Of the three new members, two of the teachers are new to the school, and one of the teachers has been teaching for over twenty years.	The four teachers in the grade 7 team have been engaged in the PLC process for four years. The team has two new team members who previously taught grade 8.
The team meets weekly during its common planning period. Team members access the agendas on their shared Google Drive.	The team meets weekly during its common planning period. Team members access the agendas on their shared Google Drive. The leader has created a focus for each meeting during the first four weeks to ensure that team members come to agreement on the essential learning standards and the common scoring agreements for the homework and common assessments the team will use throughout the unit.
At the first team meeting, the team leader shares access to the team Google Drive folder and establishes norms and a common purpose. After the first unit assessment, the team leader asks team members to bring in their student assessments and engages them in a conversation about the results.	At the first team meeting, the team leader shares access to the team Google Drive folder and establishes norms and common purpose. After the first unit assessment, the team leader asks team members to bring in their student assessments and engages them in a conversation about the results.
The two newest teachers bring in their results and begin to share their frustrations in student behaviors as a reason why their students do not perform as well as the team leader's students. The veteran teacher shares that her students did not take the assessment yet because they hadn't finished the first unit, and she won't use the assessment until all of her students understand fractions.	As they analyze student results, the team leader asks team members to complete a table tallying the results by target to find out how many students meet proficiency by target. The team notices that there is one concept on which the former grade 8 teachers' students outperform the other students.
The team leader acknowledges the team members' frustrations and tells the team to continue to move forward with the next unit, and they will discuss classroom-management issues at the next meeting.	The team analyzes the students' misconceptions, and the grade 8 teachers share a new instructional approach they think would be a re-engagement strategy for the students who did not meet proficiency. The team then puts together an intervention plan that it can use during its homeroom periods.

Figure 3.1: Grades 5 and 7 mathematics team scenarios.

LEADER *Reflection*

Both grade-level mathematics teams in figure 3.1 engage in some form of collaboration.

If you were the grade 5 team leader or another leader working with the team, what would be your next step to move the team past its current level of collaboration?

If you were the grade 7 team leader or another leader working with the team, what would be your next step to engage in deeper professional learning for all of the team members?

_____
_____
_____
_____
_____
_____
_____
_____
_____

Team Leader	Coach	Site-Level Leader
• How well does your mathematics team collaborate to answer the four critical questions? • What are your mathematics team's strengths? • What are the current challenges that you and your mathematics team need to overcome? • What structures do you currently employ to reflect on evidence of student learning? • How does your mathematics team use evidence of student learning to adapt or improve instruction? • What is the instructional focus you are currently trying to improve? • How do you make instructional practices public to learn from your mathematics team?	• Which mathematics teams exhibit high levels of collaboration and effectively answer the four critical questions? What is the evidence to support your belief? • What are the strengths of your highly effective mathematics teams, and how do you know? • What are the challenges of your ineffective mathematics teams, and how do you know? • What structures do mathematics teams use to analyze student learning? How do I support the structures? • How do the mathematics teams use the analysis of student learning to adapt or improve instruction? • What structures do mathematics teams use to make instruction transparent for continual learning?	• Which mathematics teams exhibit high levels of collaboration and effectively answer the four critical questions? • What are the strengths of your highly effective mathematics teams, and how do you know? • What are the challenges of your ineffective mathematics teams, and how do you know? • What structures do mathematics teams use to analyze student learning? How do I monitor and support the structures? • How do the mathematics teams use the analysis of student learning to adapt or improve instruction? • What structures do mathematics teams use to make instruction transparent for continual learning?

Figure 3.2: Team discussion tool—Leadership reflection on PLC structures and protocols.

Visit go.SolutionTree.com/MathematicsatWork for a free reproducible version of this figure.

What is your total score on the rubric? In which of these criteria are your teams strongest? Which present challenges?

These specific protocols when used with your teams will support stronger team engagement and a more effective response to student learning.

In chapters 4 and 5 you will examine each of the protocols your mathematics teams can use in more depth. First, however, you explore the three phases of collaboration, which will help you move your teams forward in their collaborative process.

Phases of Collaboration

Leading effective teacher collaboration (see figure 3.4 on page 65) is an essential aspect of a PLC; however, what you might consider collaboration might actually be cooperation or coordination (Grover, 1996). Initially, team meetings tend to focus more on *cooperation*. Cooperation means that team members are working first on getting to know one another while sharing some of their favorite instructional or assessment ideas. This phase requires you to support building

a community of learners and establish a trusting environment by using the leadership practices described in chapter 1 (page 11).

As you lead teams toward creating common calendars and other artifacts, they enter the *coordination phase*. During the coordination phase, teams collectively make decisions about proficiency and begin to explore evidence of student learning. Your mathematics teams pay attention to common expectations and establish a framework for equity with assessment and instructional decisions.

To move from coordination to effective collaboration, your mathematics teams look at evidence of student learning, reflect on the levels of student learning, set goals based on that analysis, and adapt instruction to better meet the needs of each and every learner in their grade level or course. In short, you help your teams to collectively own the learning of all students and not just the ones assigned to them as individuals.

Parry Graham and Bill Ferriter (2008) offer the useful framework in figure 3.4 (page 65) that details the three phases of collaboration—(1) cooperation,

Leading a Team Culture of Reflection, Refinement, and Action	Description of Level 1	Requirements of the Indicator Are Not Present	Limited Requirements of This Indicator Are Present	Substantially Meets the Requirements of the Indicator	Fully Achieves the Requirements of the Indicator	Description of Level 4
A Culture of Effective Collaboration	Members of the mathematics team are new to the PLC process and unsure of how to engage in effective collaboration. Mathematics team members are cooperative when working together, yet fail to use evidence of student learning to modify instruction. Mathematics team members and leaders rarely reflect on the impact of their actions on student learning. Leaders' feedback to teams is inconsistent or limited. Team members and leaders rarely work cohesively to impact mathematics teaching and learning.	1	2	3	4	Members of the mathematics team consistently use evidence of student learning to evaluate their instructional effectiveness. The team focuses on results and modifies unit assessments and instructional plans based on evidence of student learning. Members of the team consistently engage in cycles of inquiry and action research to reflect, refine, and act on their practice. Team members engage in structured reflections throughout the school year and modify action steps as needed to ensure they are meeting the needs of each and every learner. Leaders provide continual feedback to teams and support the collaborative process. Using intentional opportunities for reflective practices, leaders monitor and support in real time for every mathematics unit.
A Culture of Transparency With the Assessment Process	The mathematics team creates assessments for the end of the unit as a means to produce a grade for students. The team does not consistently evaluate the quality of the assessments or align them to the standards or with a balance of cognitive-demand-task levels.	1	2	3	4	Members of the grade-level or course-based mathematics team consider common unit assessments to be a tool to collect evidence of student learning for each essential learning standard. They consistently evaluate the assessment questions and tasks to ensure alignment to those agreed-on essential learning standards. When creating the assessment plan, team members pay attention to the balance of cognitive-demand levels, and the scoring routines used for consistency and accuracy regarding student feedback.
A Culture of Transparency With Evidence of Student Learning	Mathematics instructional practices are not a focus of continual reflection for the team members. Instruction is designed in isolation from colleagues. Team members may discuss instructional practices, but they do not collectively evaluate and modify the impact of their mathematics lessons.	1	2	3	4	Mathematics team members on a unit-by-unit basis discuss and analyze their instructional practices and measure the impact of their collective effort on student understanding and learning. Team members engage in structured inquiry by collaboratively planning tasks for the mathematics unit of study, developing instructional focus areas for essential standards, collecting and analyzing evidence of student learning for each essential learning standard of the unit, and taking collective action based on the evidence.

Figure 3.3: Team discussion tool—Leadership rubric for utilizing protocols for effective and focused collaboration.

Visit go.SolutionTree.com/MathematicsatWork for a free reproducible version of this figure.

Three Phases of Collaboration	Stages	Questions That Define This Stage
1. Cooperation	Stage 1: Filling the time	What exactly are we supposed to do?
	Stage 2: Sharing personal practice	What is everyone doing in his or her classroom?
	Stage 3: Planning, planning, planning	What should we be teaching, and how do we lighten the load for each other?
2. Coordination	Stage 4: Developing common assessments	What does proficiency with the essential learning standards look like?
	Stage 5: Analyzing student learning	Are students learning what they should be learning? Which students are proficient with the individual essential learning standards? Which students are not?
3. Collaboration	Stage 6: Adapting instruction to student needs	How can we adjust instruction to help both those students struggling and those students already exceeding expectations?
	Stage 7: Reflecting on instruction	Which instructional practices are most effective with our students? What will we do differently to ensure learning for all students?

Source: Adapted from Barnes & Toncheff, 2016; Graham & Ferriter, 2008; Kanold & Larson, 2012.

Figure 3.4: The three phases and seven stages of collaboration.

*Visit **go.SolutionTree.com/MathematicsatWork** for a free reproducible version of this figure.*

(2) coordination, and (3) collaboration—further broken down into seven stages of collaborative team development.

Refer back to the grade 5 and grade 7 mathematics team scenarios in figure 3.1 (page 62), and use the leader reflection to discuss which phase/stage described in figure 3.4 you think each of the two grade-level teams' actions represent. Describe the evidence that supports your designation.

Your professional responsibility as a leader is to identify your mathematics team stage within the learning progression of effective team collaboration. Then, when needed, identify protocols that you can use with the team to build a better understanding of the instructional practices that are most effective as well as the assessment processes that support deeper student understanding. You also provide feedback to promote deeper levels of team learning.

Both of the teams in the scenarios in figure 3.1 have established when they will meet using the guaranteed time provided during the school day.

Your teams may have a desire to use their time to produce "stuff" in lieu of doing the "heavy lifting" of teamwork— such as the important conversations about

LEADER *Reflection*

What feedback would you provide to the grade 5 mathematics team in figure 3.1 to move them forward to the next phase?

What feedback would you provide to the grade 7 mathematics team in figure 3.1 to move them forward to the next phase?

_____

_____

_____

_____

_____

_____

_____

_____

_____

_____

_____

evidence of student learning and ensuring equitable learning outcomes. As you lead teams in the collaborative process, ensure there is a balance between common artifact *product* development the team needs to do, and the *processes* for collective inquiry and action research necessary to analyze the teams' impact on student learning.

The protocols you choose for building shared knowledge and engaging your mathematics teams in deep reflection and actions based on evidence of student learning matter.

The Importance of a Protocol

The work of leading teams requires conversations about the agreed-on products, evidence of student learning, grading routines, and other potentially contentious topics to increase student achievement. If teams are to engage in action research and collective inquiry, you need to provide structure for the team conversations.

When you grade your own student work, you might quickly score the work, and observe similar patterns of student errors. When you engage a team of teachers to collaboratively look at student work and make the evidence of student learning public, teams shift from focusing on what was taught to what students actually learn.

Protocols provide a framework for the mathematics team discussions. Through these structured and collective team-meeting discussion questions, you help the mathematics team members to feel safe to ask questions of each other and gain new insight into student thinking. During the discussions, you monitor and encourage team members to listen to each other without judgment and identify teacher actions that impact deeper mathematical understanding, for both teachers and students.

The research of Ronald Gallimore and Bradley A. Ermeling (2010) on the impact of collaborative teams finds the use of formalized protocols that guide teacher discussions allows teachers to identify next steps versus prescribed actions, increasing the effectiveness of teacher teams. Teachers prefer protocols that allow them to be creative, apply their knowledge and skills, and learn from their peers.

Protocols provide structure for the processes collaborative teams navigate when answering the four critical questions of a PLC (DuFour et al., 2016) and engaging in action research. Protocol structures for evaluating the impact of team actions include four basic steps.

1. **Identify the question:** You and your team identify a question or a focus area to evaluate or engage in action research. Team members make predictions and describe what evidence they will evaluate for student learning.

2. **Collect the evidence:** The team collects evidence of student learning through common unit assessments or instructional observations on team-designed lessons.

3. **Analyze the evidence:** Team members meet to review the evidence of student learning and discuss results based on the evidence.

4. **Plan action steps:** Team members plan their next steps based on the results of the analysis and take action on those steps.

The story on page 67 from author Mona Toncheff provides an example of this type of team action research in mathematics.

If you want your mathematics teams to engage in collective inquiry and action research, you must provide a framework for rich discussion and team member reflection with subsequent action for the team.

In addition to using structured protocols for team interaction, you measure each mathematics team's readiness to engage in crucial conversations. As you explore protocols for analyzing evidence of student thinking, you will need to know team dynamics to best identify your team's entry point into each protocol.

Trust is the first leadership practice, and knowing your team and its level of trust in the PLC process is important. Consider your leadership role and reflect on the questions in figure 3.5.

If collaborative teams are new to sharing personal practice and student work, your first point of entry to using protocols is to make evidence of student learning public by engaging in the common assessment protocols described in chapter 4 (page 71). Once the members of your mathematics collaborative team are clear about common student learning expectations and are willing to learn from each other and trust one another, you can include instructional protocols to promote greater transparency with instruction decisions as described in chapter 5 (page 89).

Use the leader reflection to examine your team's transparency and protocols used.

Personal Story **MONA TONCHEFF**

After a few months of listening to and observing a team I was coaching, I noticed a strong level of trust between the team members was missing. The team had created an instructional vision, yet there were still strong variances from teacher to teacher in the classroom, and team members had never observed their peers. I knew I could make the instruction public with the use of lesson study; yet, the demonstration teacher would not trust the process, and the team members would not be able to engage in deep reflection due to the lack of trust.

I chose a less intrusive protocol, instructional rounds (discussed on page 90) to promote reflection and build trust. Every member of the team would be accountable for opening their classrooms, making their instruction public, and learning from each other.

Team Leader	Coach	Site-Level Leader
• What strategies does my mathematics team currently use to make its assessment and intervention routines public? • How does my mathematics team make its instruction and task routines public? • How do my mathematics team members and I continue to foster a community of learners, steeped in trust, by opening our classrooms for inspection?	• What strategies do mathematics teams currently use to make assessment and intervention routines public? • How do the mathematics teams I support make their instruction and task routines public? • How do I continue to foster a community of learners, steeped in trust, when mathematics team members are making their instructional practices transparent?	• What strategies do mathematics teams currently use to make assessment and intervention routines public? • How do the mathematics teams I support make their instruction and task routines public? • How do I continue to foster a community of learners, steeped in trust, when mathematics team members are making their instructional practices transparent?

Figure 3.5: Questions to consider when selecting team protocols and entry points.

*Visit **go.SolutionTree.com/MathematicsatWork** for a free reproducible version of this figure.*

LEADER *Reflection*

How do your mathematics teams make instruction, assessment, and intervention routines public? What protocols do you currently use with the team or teams to support more public discussions?

Culture of Reflection, Refinement, and Action

As the team leader, you are responsible for ensuring your teams' effective actions. As the coach, you are responsible for providing on-the-job mathematics training to teams and each team member. As the site-level leader, you are responsible for providing the supportive conditions necessary for continuous job-embedded learning steeped in reflection, refinement, and action.

As such, you monitor and provide just-in-time feedback to the grade-level or course-based mathematics team's progress. Continual monitoring from all leadership levels is vital if teaching and collaboration are to have a high impact. High-impact teaching requires you gather defensible and dependable evidence from many sources and hold collaborative discussions with colleagues and students about this evidence, making the effect of their teaching visible to themselves and to others.

Always strive to keep the focus on SMART goals. Include the goals on the top of every agenda you create. (See figure 2.19, page 49, as an example). List a time line of the action steps the team will complete at the end of every agenda. Help your mathematics teams plan for the next team meeting by making sure the original SMART goals at the start of the school year are ever-present.

You should also create intentional opportunities for reflection on evidence of student learning to meet the SMART goals. One strategy for continuously monitoring and providing feedback to the mathematics team SMART goals and action steps is the Evidence of Effectiveness Protocol (figure 3.6). You and your mathematics teams can utilize the protocol when reviewing the evidence of effectiveness for student learning based on current grade feedback. The instructions for the protocol appear in figure 3.7 (page 70).

The Evidence of Effectiveness Protocol (figure 3.6), also presents an opportunity for you and your team to engage in critical conversations about equitable scoring. How do you know that your team's grades are reliable? Is a C in one teacher's class describing the same level of student understanding as a C in another teacher's class? For additional support on leading this conversation, refer to *Mathematics Homework and Grading in a PLC at Work* (Kanold, Barnes, et al., in press).

LEADER *Reflection*

As you reflect on your team's progress using the Evidence of Effectiveness Protocol in figure 3.6, consider current team progress in comparison to four or five months ago.

What are the stepping-stones you are trying to reach along the way to your SMART goal? How can the tool be used to monitor both short-term (unit by unit) and long-term (state or district level) SMART goals that might not be A, B, or C grade based?

_____

_____

_____

_____

_____

_____

_____

_____

_____

_____

_____

_____

_____

_____

_____

_____

_____

_____

_____

_____

Evidence of Effectiveness Protocol

Directions: Answer each of the following grade-distribution-rate questions.

Team grade level or course: _____ Team leader: _____

1. List the current grade-distribution-rate data for each member of your team. List numbers and rates as a percentage of each team member's total.

Number of Students With Each Grade

Teacher	A	B	C	D or F
Team Total				

Percentage of Students With Each Grade

Teacher	A	B	C	D or F
Team Percentage				

2. What do the current grade-distribution-rate data reveal about overall student performance across your team?

3. Which teaching strategies helped your team achieve current student successes (As, Bs, or Cs)?

4. What do the current grade-distribution-rate data reveal about overall student performance that is not successful (Ds and Fs)?

5. What teaching strategies must your team improve or change for those students who are still struggling? How will you shift your instruction and practice to impact student improvement and move it up at least one grade?

Figure 3.6: Team discussion tool—Evidence of Effectiveness Protocol.

Visit go.SolutionTree.com/MathematicsatWork for a free reproducible version of this figure.

You and your team can use the Evidence of Effectiveness Protocol (figure 3.6) multiple times during a given grading period. Teams can use it after a unit assessment and after each grading period, whether that be quarters, trimesters, six-week sessions, semesters, or something else. The team can also adapt the protocol to monitor proficiency marks. However, the protocol is more complex when using standards-based reporting as the team will have to look at the percentage of students proficient with each standard or the overall percentage proficient with all standards in the grading time frame the team is analyzing. The protocols provided to you in chapter 4 will be better suited for analysis of proficiency on individual essential standards.

Protocol Steps (It will take about sixty minutes to complete.)	Directions
1. Identify the focus area or the question that the mathematics team is trying to answer, describe what evidence the team is evaluating, and list the predictions.	Use this protocol to answer the following questions. • What is your team's progress on meeting the SMART goal? • What percentage of the team's students is currently on target to meet the SMART goal? • What percentage of the team's students is not on target to meet the SMART goal? • Which core instruction strategies (Tier 1) are impacting student learning? • Does your team need to provide Tier 2 interventions to improve student understanding?
2. Collect the evidence.	Do the following. • Have each team member bring his or her current compiled grade-distribution-rate data or other student learning data to a team meeting. • As a team, compile these data during and at the end of each mathematics unit.
3. Analyze the evidence.	• Review the data trends and discuss the reflection questions.
4. Plan action steps based on the results of the analysis.	Do the following. • Identify Tier 1 instructional strategies to continue as a team. • Identify instructional strategies to modify. • Identify Tier 2 intervention needs and team action steps to increase the percentage of student success.

Figure 3.7: Evidence of Effectiveness Protocol instructions.

Visit **go.SolutionTree.com/MathematicsatWork** *for a free reproducible version of this figure.*

Engaging in reflection using the Evidence of Effectiveness Protocol allows teacher teams to analyze the gaps between their current reality and the SMART goal. Teams can use the protocol to reflect on core instructional strategies (Tier 1) to continue as a team and instructional strategies to modify for the upcoming unit. Teams can also plan for additional intervention needs (Tier 2) and create team action steps to increase the percentage of student success.

Looking at what students are or are not learning is less intrusive than having your peers come observe you in action. Since building trust is a critical leadership practice, encouraging mathematics teams to analyze student work is a first step in supporting a collaborative culture steeped in trust.

Next, you will explore specific protocols to make evidence of student learning become a more transparent activity among the mathematics teams you lead.

CHAPTER 4

How to Lead a Culture of Transparency and Learning With Mathematics Assessments

The most powerful lever for changing professional practice is concrete evidence of irrefutably better results.

—*Richard DuFour*

Cultivating collaboration with a culture of transparency by analyzing student learning is vital to improving mathematics teaching practices. Sharon Feiman-Nemser (1983, 2012) concludes that if schools are to produce deeper learning for students, teachers need to receive extensive and ongoing professional learning experiences.

Author and former educator Roland S. Barth (2001) says it best:

> Ultimately there are two kinds of schools: learning enriched schools and learning impoverished schools. I have yet to see a school where the learning curves . . . of the adults were steep upward and those of the students were not. Teachers and students go hand in hand as learners—or they don't go at all. (p. 23)

Creating opportunities for your team to practice careful examination of student learning opens the door for your teams' members to take risks and improve their teaching practices—and become a learning-enriched school.

Remember, the first step of all protocols is to identify the focus area or the question that your teams are trying to answer and describe the evidence to be evaluated.

The following are a few questions you might use when analyzing and learning from common mathematics unit assessment results.

- Does the common assessment we created accurately align to the unit's essential learning standards?

- Does the common assessment have a balance of lower- and higher-level-cognitive-demand tasks?

- How did each and every student perform on the common assessment by essential learning standard, and was there a specific concept with which students struggled or excelled?

- As a team, which concepts did the students learn? Which students need additional time and support to learn specific essential standards?

- How did the results vary across the team? Which instructional practices might have influenced the results? What areas warrant attention from our collaborative team or from individual team members?

This chapter addresses these questions with several protocols for analyzing evidence of student learning by allowing your teams' common mathematics unit assessment routines to become more transparent.

Figure 4.1 is a summary of the protocols in this chapter. As you review figure 4.1, think about which protocol might be an entry point into creating a culture of transparency with assessment for each mathematics grade-level or course-based team you lead.

In each section that follows, the protocol appears along with instructions on how to utilize it with additional action research questions that teams can address when engaging in a particular protocol.

The action-research question collaborative teams explore dictates the type of protocol to choose. In addition, collaborative teams benefit from a deep analysis of both the assessment tool and the evidence of student learning gleaned from the assessment.

Quality of the Common Unit Assessment Tools

How do you lead a team to create a common unit mathematics assessment? Do you use an assessment plan to create the artifact? Does your team use figure 2.2, the Team Discussion Tool: High-Quality Assessment Evaluation in *Mathematics Assessment and Intervention in a PLC at Work* (Kanold, Schuhl, et al., 2018), to evaluate the assessment tool? (Visit **go.Solution Tree.com/MathematicsatWork** to download this free reproducible.)

When leading common mathematics unit assessment development, a clearly articulated assessment plan supports your collaborative teams to make informed decisions when choosing tasks that align to each standard and to format the assessment to efficiently analyze the evidence of student learning it produces.

Common Assessment Analysis Protocol

If you think about a current common unit mathematics assessment you helped a team create, do you know how many questions align to each essential learning standard? Do you know how many *higher-* or *lower-level-cognitive-demand* test questions align to

Actions If you want to:	Protocol	Questions the Protocol Will Answer	Figure and Page Number
Analyze the quality of your common unit mathematics assessments and alignment to the essential learning standards of the unit.	Common Mathematics Unit Assessment Analysis Protocol	Do all team members have an understanding of the assessment evidence the assessment tool is gathering?	Figure 4.2
Analyze the learning differentials between team members by essential standard or by learning target.	Essential Standards and Target Proficiency Protocol	Do all team members know in which essential standards the students are or are not yet proficient?	Figure 4.4 (page 75)
Analyze specific learning trends at various levels of proficiency and create plans for student re-engagement by essential standard as needed.	Looking at Student Work Protocol	Do all team members have agreement regarding student work between the levels of *below, approaching, meeting,* or *above* proficiency?	Figure 4.6 (page 77)
Analyze proficiency levels of students by each essential learning standard for the mathematics unit.	Essential Learning Standard–Analysis Protocol	Do all team members know how many students are proficient with each essential learning standard for the unit? Do teachers know the instructional strategies that support proficient or not-yet-proficient student work?	Figure 4.8 (pages 80–81)
Analyze the complexity of thinking or the solution pathways students take to problem solve and reason mathematically.	Student Thinking and Reasoning Protocol	Do all team members know different problem-solving strategies and mathematical representations students should use?	Figure 4.10 (pages 85–86)

Figure 4.1: Summary of protocols for building a culture of transparency with assessment.

each essential learning standard? Does everyone on the team agree on the scoring to be used when grading each task on the common assessment?

When you answer these questions, consider using figure 4.2, the Common Mathematics Unit Assessment Analysis Protocol, to evaluate team-created common assessments. Figure 4.3 contains the directions for using the protocol with your team. Read the protocol instructions, and use the protocol to consider your current assessments.

Common Mathematics Unit Assessment Analysis Protocol				
Essential Standards	Assessment Item or Items	Cognitive-Demand Balance	Points or Rubric Score	Percentage of Test

Consider the following items.
- Does the percentage of the test devoted to each target align to the expected cognitive demand assessed?
- Does the number of items that align to each essential standard align to the instructional time spent on the essential standard?
- Should we revise the assessment? If so, how?
- Based on the answers to these questions, should we revise the instructional plan? If so, how?

Figure 4.2: Common Mathematics Unit Assessment Analysis Protocol.

Visit go.SolutionTree.com/MathematicsatWork for a free reproducible version of this figure.

Protocol Steps (It should take about sixty minutes to complete the following four steps.)	Directions
1. Identify the focus area or the question that the mathematics team is trying to answer, describe what evidence the team will evaluate, and list the predictions.	This protocol will help your team answer the following three questions. 1. How many questions align to the essential learning standard? 2. How many higher- or lower-level-cognitive-demand questions align to each essential learning standard? 3. Does everyone on the team agree on the scoring of each item?
2. Collect the evidence.	Identify a current assessment to analyze. Initially, choose a common assessment that you know needs improvement or a unit assessment on which students typically underperform.
3. Analyze the evidence.	Using the Common Mathematics Unit Assessment Analysis Protocol, complete the analysis of the common assessment and reflection questions.
4. Plan action steps based on the results of the analysis.	Revise the assessment or the instructional plan as you need to.

Figure 4.3: Common Mathematics Unit Assessment Analysis Protocol instructions.

Visit go.SolutionTree.com/MathematicsatWork for a free reproducible version of this figure.

Using the protocol directions from figure 4.3 (page 73) as a guide, review figure 4.2 (page 73) and identify how you would use this protocol to support the development and use of current common mathematics assessments.

The Common Mathematics Unit Assessment Analysis Protocol described in figure 4.2 will assist your team assessment conversations. If your teams are not currently using common assessments or discussing student work together, the protocol creates a process for developing trust and agreement. When using the protocol, your teams are analyzing a current unit mathematics assessment and making sense of the essential learning standards for the unit at the same time. Take a moment to reflect on how you could use this protocol with your team to build shared knowledge about high-quality and common assessments.

LEADER *Reflection*

How could you use the Common Mathematics Unit Assessment Analysis Protocol with your team to build shared knowledge of best practices with common mathematics unit assessment design?

_____

_____

_____

_____

_____

_____

_____

_____

When you engage your grade-level or course-based teams in a common assessment analysis, you are building capacity by ensuring team members possess a strong understanding of the essential learning standards for the unit and the expectations for evidence of student learning and instruction for those standards. You also establish equitable scoring expectations to address potential scoring/grading inequities (see Kanold, Schuhl, et al., 2018, in *Mathematics Assessment and Intervention in a PLC at Work,* 2018).

Once you and your team are confident with the quality of your common unit mathematics assessments, you are ready to engage your teams in analyzing evidence of student learning— using the next set of protocols.

Evidence of Student Learning Protocols

When analyzing evidence of student learning, teams take actions to close learning gaps with students who are not yet meeting proficiency as well as provide enrichment for students who are ready to dig deeper in the content. If your collaborative teams only participate in common assessment planning, your team will fail to fully engage in answering the third and fourth critical questions of a PLC (DuFour et al., 2016)—(3) How will we respond when some students do not learn?, and (4) How will we extend the learning for students who are already proficient?

When evaluating student work, your teams should do the following.

- Develop and adhere to norms for analyzing student work.

- Complete student tasks individually before sorting student work. Knowing the cognitive demands of tasks encourages greater in-depth analysis and understanding of student thinking.

- Work collaboratively to identify the root cause of the strengths or challenges in student understanding.

When you lead your teams to uncover root causes, it is important to focus on students' mathematical thinking versus student effort. It is easy to blame poor student performance on students' attendance or failure to complete homework. However, to create a targeted response to student learning, lead your teams to dig deeper and discuss the academic understanding students show in their work and the implications for their future work.

Essential Standards and Target Proficiency Protocol

The Essential Standards and Target Proficiency Protocol allows collaborative team members to share their data by identifying the number of students who are proficient by essential standard and target. It is a first step in data analysis to promote transparency with assessment evidence.

The evidence of learning in this tool is limited to identifying trends and patterns across the team. For some teachers, sharing results of student learning (or lack thereof) can be difficult due to a fear of being judged by colleagues. You can minimize this fear by engaging teachers to share percentage results with their peers. As teachers examine the student results for all members of the team, you can lead focused discussions related to student learning—both about what students learned or what they have not learned *yet*—as well as discussions on the instructional practices that may have led to the results. Figure 4.4 shows a completed protocol from an algebra 1 course unit of study to illustrate the information teachers can collect and analyze. The percentage of students proficient on each essential standard is shared, and then the percentage of each

individual learning target can also be analyzed for a deeper analysis of learning trends. The directions for the protocol appear in figure 4.5 (page 76).

If this were your team, what questions would you ask your team members? What suggestions or recommendations would you provide to your team?

Teachers can collect their student performance individually within a few days of giving the common mathematics assessment and bring the collected evidence to the team meeting. You can use Google Docs (https://docs.google.com) or create a form to easily collect the team results. Once the team collects the percentages by essential learning standard, the team can review the results and discuss next steps during a team meeting. Use the leader reflection on page 76 to consider your team's readiness for publicly sharing results.

Unit 1: Expressions	Teacher			
	1	2	3	4
Essential standard 1: Reason quantitatively and use units to solve problems.	**61 percent**	**40 percent**	**65 percent**	**70 percent**
I can use units as a way to understand problems and to guide the solution of multistep problems; choose and interpret units consistently in formulas; choose and interpret the scale and the origin in graphs and data displays.	56 percent	45 percent	67 percent	70 percent
I can define appropriate quantities for the purpose of descriptive modeling.	55 percent	39 percent	66 percent	69 percent
I can choose a level of accuracy appropriate to limitations on measurement when reporting quantities.	77 percent	50 percent	63 percent	71 percent
Essential standard 2: Use properties of rational and irrational numbers.	**77 percent**	**37 percent**	**37 percent**	**58 percent**
I can explain why the sum or product of two rational numbers is rational; that the sum of a rational number and an irrational number is irrational; and that the product of a nonzero rational number and an irrational number is irrational.	77 percent	37 percent	37 percent	58 percent
Essential standard 3: Interpret the structure of expressions.	**61 percent**	**38 percent**	**70 percent**	**75 percent**
I can interpret expressions that represent a quantity in terms of its context, and I can interpret complicated expressions by viewing one or more of their parts as a single entity.	61 percent	38 percent	70 percent	75 percent

Figure 4.4: Essential Standards and Target Proficiency Protocol—Algebra 1 example.

*Visit **go.SolutionTree.com/MathematicsatWork** for a free reproducible version of this figure.*

Protocol Steps (It should take about fifty minutes to complete the following four steps.)	Directions
1. Identify the focus area or the question that the team is trying to answer, describe evidence the team will be evaluating, and list the predictions.	This protocol will help your team answer the following three questions. 1. What percent of the students are proficient on each essential learning standard for the unit? 2. What percent of the students are proficient on the daily targets tied to each essential standard? 3. Which essential learning standards do we need to spend more time on as a team versus standards that students are mastering?
2. Collect the evidence.	Identify the common assessment you will analyze as a team. Be sure the questions align to the essential learning standards prior to compiling the student results. Create a target tracker or utilize a data-management tool to collect the evidence. Team members should bring their individual data to the team meeting and be prepared for the discussion. Note: Before using this protocol, team members come to agreement on the scoring of each essential learning standard or target and expectations of proficiency.
3. Analyze the evidence.	Using the Essential Standards and Target Proficiency Protocol, as a team, complete the analysis of what percentage of students is meeting proficiency with each target. Next, identify trends in student performance across the team and what content students have mastered. Discuss similarities and differences. Identify possible causes, and brainstorm next steps.
4. Plan action steps based on the results of the analysis.	List two or three action steps based on evidence of learning. Identify content needs and areas for enrichment. Discuss different instructional strategies that the team can use to increase student learning.

Figure 4.5: Essential Standards and Target Proficiency Protocol instructions.

*Visit **go.SolutionTree.com/MathematicsatWork** for a free reproducible version of this figure.*

LEADER *Reflection*

What is your team's readiness for collectively sharing unit assessment results publicly? What action steps would your team need to complete to use the Essential Standards and Target Proficiency Protocol?

_____

_____

_____

_____

_____

_____

_____

_____

A possible limitation of the protocol in figure 4.4 (page 75) is that teams focus on the percentage of students meeting proficiency instead of having a deeper conversation about student thinking; the mathematics team conversations are surface level. However, this is a first step for analyzing student work, and it provides a starting point for team members to share their current reality.

As you lead teams though this protocol or one of similar design for your grade level or team, you can begin discussions related to instructional practices that may have been more effective (or not) toward helping students learn specific concepts. For example, after reviewing the trends of student performance in figure 4.4, the algebra team identifies two essential learning standards that were challenging for students. The team is able to discuss possible challenges, address next steps for re-engagement, and create a tiered response to student learning.

Some of the actions that the team might take include the following.

- Use a buffer or flex day in the schedule to address student misconceptions during small-group instruction.

- Determine learning targets to address during an intervention time in the school schedule; identify a new instructional approach and determine which teacher will support students for each learning standard.

- Share students during core instruction or intervention time; for example, students who need support with the first learning target go to teacher A, second learning target to teacher B, and so on.

- Add additional warm-up activities for two weeks to see if that strategy makes a difference in learning.

A natural progression is for your team to identify the challenging standard areas and then transition to sorting student work based on proficiency levels, analyze the actual student thinking to understand student errors and their current level of mastery, and provide time to improve.

Looking at Student Work Protocol

The Looking at Student Work Protocol (figure 4.6) is the next step to a deeper analysis of student learning. You will lead your team to develop clear expectations for four different levels of performance on either a common mid-unit or unit mathematics assessment.

You can use this protocol with one class set of student papers from a designated teacher, or each teacher can bring in one class set of student work. There are benefits when you ask each team member to bring in his or her

Looking at Student Work Protocol			
Teacher name: _____ Team: _____			
Period: _____ Date: _____			
Student work for analysis: _____			
Learning targets or essential learning standards: _____			
1. Analyze the expectation for student work or performance at each level—What are the criteria to assess this work?			
Below Standard	**Approaching Standard**	**Meeting Standard**	**Exceeding Standard**
2. Sort student work and list student names at the appropriate level. Identify what percentage of students are at each level.			
Below Standard	**Approaching Standard**	**Meeting Standard**	**Exceeding Standard**
_____ percent of class	_____ percent of class	_____ percent of class	_____ percent of class

3. Choose one or two students from each level, and describe the evidence of student learning. Focus on the following four areas.
    a. Demonstrates deep conceptual understanding
    b. Shows procedural knowledge of mathematical content
    c. Demonstrates skills and understanding in problem solving
    d. Demonstrates effective communication			

Below Standard	**Approaching Standard**	**Meeting Standard**	**Exceeding Standard**
4. What instructional changes do we need to make?			
Below Standard	**Approaching Standard**	**Meeting Standard**	**Exceeding Standard**

Figure 4.6: Looking at Student Work Protocol.

Visit go.SolutionTree.com/MathematicsatWork for a free reproducible version of this figure.

Personal Story **MONA TONCHEFF**

When I used the Looking at Student Work Protocol with an algebra 1 team, members noticed students were struggling with using the quadratic formula. As the teams sorted student work, I overheard team members stating they would need to reteach everything about the quadratic formula.

After inspection of the student work during the protocol, the team realized that students actually understood the formula and when to use the formula. However, when using the formula, students were making mistakes with the calculations related to $\sqrt{b^2 - 4ac}$. The teachers were able to create a plan for students to re-engage with the procedural aspect of simplifying expressions involving square roots, squares, and negatives, and they avoided spending an entire day reteaching a concept that students actually understood.

own set of student work, especially if teachers are new to *data dialogues*.

By asking each team member to bring in student work, you embed teacher ownership and responsibility for accurately scoring and grading student work for the agreed-on common assessment. You are also modeling how to honor the team's commitment to the PLC process and the four critical questions.

Additionally, when a team member is looking at his or her student work only, the risk taking involved in making his or her student thinking public is lower. This is another opportunity to build confidence in the process. Another benefit of the Looking at Student Work Protocol is that teams dig deeper into what students do or do not yet understand.

See figure 4.7 to read the directions for the Looking at Student Work Protocol. Think about an upcoming assessment and how this protocol would support one of your team goals.

Think about the similarities and differences between this protocol and the previous Essential Standards and Target Proficiency Protocol in figure 4.4 (page 75). The

Essential Standards and Target Proficiency Protocol limits the review of student thinking into two categories: Proficient or not proficient. The Looking at Student Work Protocol has teams analyze student learning as a progression of developing proficiency and extending proficiency. This deeper analysis promotes richer discussions moving past correct versus incorrect student work.

When your team is analyzing student work, it has intentional opportunities to identify specific mathematics learning gaps and strengths. When teams are completing the third step of the protocol (analyze the evidence) in figure 4.7, a natural result of analyzing misconceptions is for team members to share instructional approaches or strategies that impacted student learning. This, in turn, builds shared knowledge of effective practices. Use the leader reflection to take a few moments to reflect on how this protocol supports team exploration of student thinking.

As you and your team use the Looking at Student Work Protocol, your team discussions and depth of analysis will improve each time. This protocol is a first step for helping your teams learn how to analyze student thinking.

LEADER *Reflection*

For the mathematics grade-level or course-based teams you support, are there specific mathematical content standards that students continue to struggle with learning each unit and each year?

How might this protocol support your teams' exploration of improved strategies for student learning?

_____

_____

_____

_____

_____

_____

_____

Protocol Steps (The following four steps take about sixty minutes to complete.)	Directions
1. Identify the focus area or question the mathematics team is trying to answer, describe evidence the team will evaluate, and list the predictions.	This protocol will help your team answer the following questions. 1. What is the difference in student work between the rubric levels *below*, *approaching*, *at*, or *above* proficiency? 2. What percentage of the students are at each level in individual classes and across the team? 3. What specific misconceptions can you address that may be across all four levels of proficiency? 4. Which instructional strategies are working? Which are not working? 5. Are there re-engagement or enrichment strategies your teams collectively can use?
2. Collect the evidence.	Agree on the common assessment to analyze. Team members should bring their individual data to the team meeting and be prepared for the discussion. Note: Before using this protocol, teams come to agreement on the scoring of each essential learning standard and target and the definitions of the four levels of proficiency.
3. Analyze the evidence.	If the team has not done so already, members begin by defining the four levels of proficiency. Each teacher sorts one class of scored student work, or one teacher can share his or her student work and the team can complete the protocol for one class together. Once sorting of student work is complete, list the names of each student at each level and calculate the percentage of students in each level. Circle one or two student names from each level and then analyze and summarize their performance. List observations and share results across the team.
4. Plan action steps based on the results of the analysis.	List two to three action steps based on evidence of learning. Identify content needs and areas for enrichment. List different instructional strategies to utilize for re-engagement or enrichment.

Figure 4.7: Looking at Student Work Protocol instructions.

Visit go.SolutionTree.com/MathematicsatWork for a free reproducible version of this figure.

Essential Learning Standard Analysis Protocol

The next protocol for analyzing assessment evidence, the Essential Learning Standard Analysis Protocol, structures in-depth mathematics team conversations around evidence of student thinking aligned to the essential learning standards and targets of the assessments.

As part of the collective response to student learning, you and your team are expected to create a tiered, required response to student learning. The Essential Learning Standard Analysis Protocol provides a framework to articulate the mathematics content standards students need for additional support and extension based on evidence of student learning. Figure 4.8 (pages 80–81) shows an example of a *grade 4 team's* analysis of a unit assessment related to fractions. (Visit **go.SolutionTree.com/MathematicsatWork** to find a free reproducible blank template for figure 4.8.)

Read the instructions for the Essential Learning Standard Analysis Protocol in figure 4.9 (page 82). Think about an upcoming assessment and how this protocol could be used to support the professional work of your teams. (Visit **go.SolutionTree.com/MathematicsatWork** to find a free reproducible version for figure 4.9.)

Essential Learning Standard Analysis Protocol

1. Define each essential standard on the assessment and describe the expectation of proficiency. Write your definitions and descriptions in the following chart.

	Essential Standard 1	Essential Standard 2	Essential Standard 3	Essential Standard 4
Expectations of Proficiency	I can explain why fractions are equivalent and create equivalent fractions. Students can explain why a fraction is equivalent to another fraction by using multiple representations. They will also describe how the number and size of the parts differ even though the two fractions themselves are the same size.	I can compare two fractions and explain my thinking. Students will compare the two fractions with and without common denominators and explain how to compare two fractions with different denominators. They will also justify their comparison using a model.	I can add and subtract fractions, show my thinking, and use one or more models to justify my response. Students are able to solve the problems (with no errors) and use one or more models to justify their response.	I can multiply a fraction by a whole number and explain my thinking. Students are able to multiply a whole number and a fraction (no errors) and use one or more models to justify their response.

2. Determine the number and percentage of students proficient on the assessment for each standard by teacher and then for all students within the team. Write the information in the following chart.

	Essential Standard 1		Essential Standard 2		Essential Standard 3		Essential Standard 4		Total Number of Students
	Number	Percent	Number	Percent	Number	Percent	Number	Percent	
Teacher A	20	65	10	32	22	70	28	90	31
Teacher B	19	68	12	43	20	71	26	92	28
Teacher C	20	65	8	29	28	90	25	80	31
Total Team	59	66	30	33	70	78	79	88	90

3. For each standard, determine which students are unsatisfactory, which have limited knowledge, which are proficient, and which are advanced by teacher and as a team.

Essential Standard 1					
	Unsatisfactory	Limited Knowledge	Proficient	Advanced	Total Number of Students
Teacher A	2	9	10	10	31
Teacher B	8	1	19	0	28
Teacher C	11	0	16	4	31
Total Team	21	10	45	14	90

Essential Standard 2					
	Unsatisfactory	Limited Knowledge	Proficient	Advanced	Total Number of Students
Teacher A	10	11	2	8	31
Teacher B	10	6	12	0	28
Teacher C	19	4	4	4	31
Total Team	39	21	18	12	90

Essential Standard 3					
	Unsatisfactory	Limited Knowledge	Proficient	Advanced	Total Number of Students
Teacher A	4	5	7	15	31
Teacher B	7	1	19	1	28
Teacher C	3	0	18	10	31
Total Team	14	6	44	26	90

Essential Standard 4					
	Unsatisfactory	Limited Knowledge	Proficient	Advanced	Total Number of Students
Teacher A	1	2	20	8	31
Teacher B	0	2	22	4	28
Teacher C	6	0	24	1	31
Total Team	7	4	66	13	90

4. Which essential standards were student strengths? What instructional strategies impacted student thinking?

Our students are doing well with adding and subtracting common denominators and multiplying fractions by a whole number. Having the students engage in number talks during this unit has really helped students make connections between the models that students create and the thought process.

5. In which areas did individual teachers' students struggle? In which areas did our team's students struggle? What is the cause? How will we respond?

Only 66 percent of our students are proficient with equivalent fractions. Teacher A has been using more manipulatives and will use them with students from teachers B and C.

As a team, students are struggling with comparing two fractions when the denominator is not common. They are confusing the models or are not being precise when they use a circle when comparing fractions. We will create a plan for the students who need more time and support to include a focus on fraction representation using the rectangular model and the applet from NCTM.

6. Which students need additional time and support to learn the standards? What is our plan?

Next week during intervention time, we will use the following schedule:

Monday and Tuesday—Teacher A and support staff will work with the thirty-nine identified students on comparing two fractions. Teachers B and C will use the recipe task to stretch students' understanding of uncommon denominators.

We will also use small-group instruction and centers during the week to do more work with manipulatives. Teacher A is going to bring her manipulatives to share at the next team meeting.

7. Which students need extension or enrichment? What is our plan?

See notes on the use of the recipe task.

Source: Adapted from Kramer & Schuhl, 2017.

Figure 4.8: Essential Learning Standard Analysis Protocol—Grade 4 sample.

Visit go.SolutionTree.com/MathematicsatWork for a free reproducible version of this figure.

Protocol Step (The following four steps should take about ninety minutes to complete.)	Directions
1. Identify the focus area or question the mathematics team is trying to answer, describe evidence the team will use to evaluate student performance, and list predictions for that performance.	This protocol will help your team answer the following seven questions. 1. How many students are unsatisfactory, have limited knowledge, are proficient, and are advanced by teacher and as a team? 2. What specific misconceptions can you address that may be across all four levels of proficiency? 3. Which instructional strategies are working? Which are not working? 4. What re-engagement or enrichment strategies can your team collectively use? 5. In which areas did individual teachers' students struggle? In which areas did your team's students struggle? What is the cause? How will you respond? 6. Which students need additional time and support to learn the standards? What is your plan? 7. Which students need extension or enrichment? What is your plan?
2. Collect the evidence.	Note: Before the team uses this protocol, team members need to reach agreement on the scoring of the math tasks for each essential learning standard and expectations of proficiency for those standards. Agree on the common mathematics unit assessment to analyze. Be sure the questions align to the essential standards prior to compiling the student results. The team members will bring a class set of student work to the team meeting and will be prepared for the discussion.
3. Analyze the evidence.	The team will define the four levels of proficiency. Each teacher will sort one class of student work, or one teacher can share his or her student work and the team can complete one class together. Once the work is sorted, list the number of students at each level and calculate the total for the team. Complete the team reflection questions.
4. Plan action steps based on the results of the analysis.	List two to three action steps based on evidence of learning. Identify content needs and areas for enrichment. Create additional interventions as needed to address the learning needs.

Figure 4.9: Essential Learning Standard Analysis Protocol instructions.

*Visit **go.SolutionTree.com/MathematicsatWork** for a free reproducible version of this figure.*

When you lead your grade-level or course-based collaborative teams through an analysis of student learning, standard by standard, your teams will develop a better understanding of the proficiency levels for each essential standard. Use the following leader reflection to consider when you could use the protocol presented in figure 4.8 (pages 80–81) to support the assessment work of your team.

As a result of using the Essential Learning Standard Analysis Protocol, your mathematics team will be able to identify specific students who meet proficiency and

LEADER *Reflection*

How can you use the Essential Learning Standard Analysis Protocol (figure 4.8) to support your team's understanding of student learning for each unit of mathematics?

_____

_____

_____

_____

_____

_____

target those students who yet need additional time and support for each mathematics standard.

This level of data dialogue informs the collective *team* response to student learning. Using a Tier 2 type of intervention plan (similar to those described in *Mathematics Assessment and Intervention in a PLC at Work* (Kanold, Schuhl, et al., 2018), your team ensures greater equity in student learning outcomes across all teachers.

Tier 2 mathematics interventions target students in your grade level or course not yet meeting proficiency for the standards in a given mathematics unit and in need of additional time and support outside of the daily lesson. The instruction provided during any Tier 2 intervention time should target specific deficits by using the learning targets and essential learning standards from previous and current mathematics units as the focus for the intervention.

Your team-developed Tier 2 interventions should address common student errors revealed when your teams analyze student thinking on common assessments. Your team's priority is to identify meaningful instructional strategies and mathematical tasks that differ from the initial core instruction.

The team in figure 4.8 (pages 80–81) defined proficiency for each of the four essential standards in the first step and sorted the student work accordingly.

When you first consider the proficiency rate by target, the percentages for each essential learning standard across the team members are somewhat consistent. However, when you sort the student work by levels of proficiency, you and your team will be able to clearly identify areas of student need for meeting proficiency.

If the grade 4 example in figure 4.8 was your collaborative team, you might notice that one team member identifies more students at the advanced level and one team member has several students far from proficient. Discussion of the team data may reveal that one instructional strategy impacted student understanding more than another. In fact, only one teacher might have used the strategy. If that is the case, then during a Tier 2 intervention, the teacher who uses the effective strategy might work with all the students who still need to learn the specific essential learning standard during an intervention time.

Looking at student thinking will prompt even deeper discussions across the team focused on how to impact student learning. Which specific teacher actions during core instruction promote deeper student understanding, and which actions do not?

As you and your team gain more experience with protocols for analyzing student work, you will be able to modify and adjust the protocol to meet your assessment needs.

Personal Story **SARAH SCHUHL**

When I used the Essential Learning Standard Analysis Protocol (figure 4.8) with a grade 1 (first grade) team, one teacher had more proficient and advanced students than any other team member; however, when she shared her student work, the evidence of proficient and advanced was not apparent, so the team questioned her. She replied that she knew the students understood the concepts from their work in class—they just didn't show all of their thinking on the test.

We had to stop and talk about how looking at student work requires an understanding that the student thinking shown alone determines proficiency and instruction, and that teachers need to work with students to understand the expectations in writing their thinking and solutions to tasks.

Student Thinking and Reasoning Protocol

The Student Thinking and Reasoning Protocol is a final structure you can use with teams to evaluate: (1) the effectiveness of a singular mathematical task or smaller assessment and (2) the level of student thinking, solution pathways, selection of strategies, or problem solving students use to complete the task. Notice the grain size of the assessment evidence is smaller as you help teachers to focus on the mathematical thought processes for a specific learning target or essential learning standard.

When using the Student Thinking and Reasoning Protocol (figure 4.10), it is best to use a higher-level-cognitive-demand mathematical task or set of tasks that requires students to explicitly describe and justify their reasoning and allows for multiple entry points and solution pathways. If the teams you support choose lower-level-cognitive-demand mathematical tasks that limit student thinking or have only one solution pathway, discussions will not lead to a deeper understanding of student thinking and reasoning.

Figure 4.10 is a completed Student Thinking and Reasoning Protocol by a grade 2 team of teachers. With your team, read through the sample student work and reflect on the suggested additional interventions and team actions. What additional action steps might you and your team create based on the evidence of student thinking?

As you analyze the student work in figure 4.10, what do you notice about the levels of student thinking? During its analysis, the grade 2 team noticed that across all three levels, students were not using open number lines. The team was able to come up with a plan on how to teach in this strategy during the upcoming unit.

As your teams engage in deeper conversations about student thinking and the expectation of the evidence in quality of student thinking and reasoning, your team will develop a deeper understanding of the mathematics standards and a shared knowledge of the instructional practices that promote effective problem solving (see Kanold, Kanold-McIntyre, et al., 2018).

The Benefit of Protocols

When analyzing student work, you lead your team beyond the coordination and cooperation phases discussed in chapter 3 (figure 3.4, page 65) to effective collaboration that is steeped in personal and collective reflection focused on student learning results.

Your intentional actions and the use of the protocols from this chapter will promote a cycle of inquiry to help your mathematics teams address:

- How is our team identifying questions to further analyze the quality of our common unit assessments?

- What types of tasks or assessment questions can elicit more depth in student thinking?

- Which instructional practices are impacting student learning the most?

- What action steps does our team need to complete to increase student understanding?

- What re-engagement strategies should our mathematics team use for the students not yet proficient? When will the intervention take place, and how will the team structure the time?

- What enrichments must our team plan based on students whose work is above proficiency? When will it take place during the school day, and how will the team structure the time?

Engaging your mathematics teams in structured protocols analyzing both the assessment tool and the evidence of student learning increases a team's shared knowledge of both the expectations for learning the standards and the research-affirmed actions that lead to increased student understanding.

Additionally, analyzing evidence of student learning provides you with the opportunity to respond to the team's assessment-learning needs and collectively respond to the third and fourth critical questions of a PLC (DuFour et al., 2016) in relation to the teams or team members you serve.

3. How will we respond if the *team members* don't know it?

4. How will we respond if the *team members* do know it?

Evaluating Student Thinking

Part 1: Completed Rubric for the Three Levels of Student Thinking

Teacher name: Mr. Gallegos

Team: Grade 2 team

Learning target: I can solve word problems and show my thinking.

Task: David goes to the park. He sees 15 dogs, 23 squirrels, and some birds. All together he sees 52 dogs, squirrels, and birds at the park. How many birds did David see at the park? Show how you know your answer is correct.

Student Thinking Evidence	Student Names at Each Level
The following are generic levels of student thinking. **Above proficient:** • Uses an efficient and effective strategy to solve the problem • Provides a detailed and clear explanation with mathematical representations he or she used to expand the solution • Uses appropriate mathematical vocabulary • Justifies the solution pathway • Shows complete understanding of the concepts, skills, and procedures	
Proficient: • Uses an effective strategy to solve the problem, but makes minor errors • Provides a clear and supported explanation with mathematical representations • Uses some of the mathematical vocabulary • Includes some justification of the solution pathway • Shows substantial understanding of the mathematical concepts, skills, or procedures	
In progress: • Uses strategies that are not appropriate for the problem • Lacks explanation or clarity of explanation • Lacks precise mathematical language • Does not always provide clear justification for the chosen solution pathway • Shows limited understanding of the mathematical concepts, skills, or procedures	

Part 2: Agreed-On Rubric for the Three Levels of Student Thinking

Teacher name: Mr. Gallegos   **Team:** Grade 2 team

Learning target: I can solve word problems and show my thinking.

Task: David goes to the park. He sees 15 dogs, 23 squirrels, and some birds. All together he sees 52 dogs, squirrels, and birds at the park. How many birds did David see at the park? Show how you know your answer is correct.

With your team, come to agreement and define the expectation of each level. Sort the student work according to the description of each level, and list each student name accordingly.

Student Thinking Rubric (Part 1)	Student Names at Each Level (Part 2)
Above proficient: Student solves the task correctly with work to show the answer is correct and checks the work or shows two ways to justify the answer, including birds as units at the end of the answer.	Student 1 Student 2 Student 3

Figure 4.10: Student Thinking and Reasoning Protocol—Grade 2 example.

continued ↓

Proficient:

Student solves the task correctly with work to support the answer and may not have the units as part of the answer. Student may support his or her answer in more than one way. However, he or she may have one error in one of the representations.

| Student 4 |
| Student 5 |
| Student 6 |
| Student 7 |
| Student 8 |
| Student 9 |
| Student 10 |
| Student 11 |
| Student 12 |

In progress:

Student has correct work with an incorrect answer or a correct answer, but the work does not support the answer.

Or

Student shows minimal understanding in work and answer. (For example, student might solve $15 + 23 + 52 = ?$ correctly.)

| Student 13 |
| Student 14 |
| Student 15 |
| Student 16 |
| Student 17 |

Student 3 work

Student 8 work

Student 14 work

Part 3: Description of Student Thinking Evidence

Teacher name: Mr. Gallegos   **Team:** Grade 2 team

Learning target: I can solve word problems and show my thinking.

Task: David goes to the park. He sees 15 dogs, 23 squirrels, and birds at the park. All together he sees 52 dogs, squirrels, and birds at the park. How many birds did David see at the park? Show how you know your answer is correct.

With your team, come to agreement and define the expectation of each level. Sort the student work according to the description of each level, and list each student name accordingly.

Student Thinking Evidence	Next Steps for Enrichment or Re-Engagement
Above proficient	Students are using the graphic organizer effectively and are precise with the units. Provide challenge task during enrichment time with numbers larger than 100. Possibly include problems that require more than two operations.
Proficient	Continue to work with students to help them develop additional strategies. Include time during intervention for number lines and breaking apart the numbers. Create a stations activity for reinforcing multiple strategies, and, when asking students to present their work under the document camera, create a rubric for what students should have on their work.
In progress	Need to use base-ten blocks and the number line to reinforce the concept of a missing addend. Need more work on taking the given information in the task and creating a bar model to represent what information is missing. Use open number lines in the upcoming unit to reinforce addition.

Visit go.SolutionTree.com/MathematicsatWork for a free blank reproducible version of this figure.

Use the leader reflection to reflect on your role as leader in supporting your team when it is struggling or moving members forward as they excel.

LEADER *Reflection*

If the team is still struggling with establishing common scoring and expectations for student learning, what actions can you take as the team leader, coach, or site-level leader to help the team members?

If the team is excelling, what actions can you take to move the team deeper into their work?

_____

_____

_____

_____

_____

_____

_____

_____

_____

_____

_____

_____

_____

_____

_____

_____

_____

_____

_____

_____

The instructions for the Student Thinking and Reasoning Protocol appear in figure 4.11 (page 88). Think about an upcoming higher-level-cognitive-demand mathematical task for which this protocol would support one of your team goals.

When you create opportunities for careful examination of student learning, you open the door for team members to take risks and improve their teaching practices. You also ensure more equitable opportunities for students to learn or re-engage in learning, regardless of which classroom teacher of mathematics the school assigns.

As you engage your mathematics team in this work, you are helping to shift the team's focus and energy from analyzing assessment evidence of student learning to their instructional practices and the impact of those routines on daily student learning from one mathematics unit to the next.

TEAM RECOMMENDATIONS

Leading a Culture of Transparency and Learning With Assessments

- Choose protocols for team engagement based on your team's current level of trust.

- Use a protocol two or three times before moving to a new protocol so team members can engage in more in-depth conversations about student learning.

- Provide time for teacher *team reflection* on student learning with evidence from the common mathematics assessment tools.

- Work together with other leaders to monitor and support team reflection on common assessment evidence and require team action based on evidence of student learning.

Protocol Steps (The following four steps take about sixty minutes to complete.)	Directions
1. Identify the focus area or the question that the team is trying to answer, describe what evidence the team will evaluate, and list the predictions.	This protocol will help your team answer the following six questions. 1. What solution pathways are students using to solve this task or tasks? 2. What are the most effective solution pathways? 3. What problem-solving strategies are students using? 4. Which instructional strategies are working, and which should we modify to encourage more productive problem-solving strategies? 5. Which students need additional time and support to develop more meaningful strategies? 6. Which students need extension or enrichment? What is your plan?
2. Collect the evidence.	Agree on the common task or tasks to analyze. Be sure to choose a task and a focus problem-solving strategy or strategies to evaluate when compiling the student results. Team members should bring a class set of scored student work to the team meeting and be prepared for the discussion. Note: Before using this protocol, teams have to come to agreement on the scoring of each essential learning standard and target and define the three levels of proficiency.
3. Analyze the evidence.	Using the Student Thinking and Reasoning Protocol, the team will either use the generic rubric with defined levels of thinking (part 1) or define and agree on three levels of student thinking (part 2). Each teacher will sort one class of student work based on the evidence of student thinking. Once the team sorts the work, list the names of students at each level. In part 3, each team member describes the evidence of student thinking and discusses the evidence.
4. Plan action steps based on the results of the analysis.	List two or three action steps based on evidence of learning.

Figure 4.11: Student Thinking and Reasoning Protocol instructions.

*Visit **go.SolutionTree.com/MathematicsatWork** for a free reproducible version of this figure.*

In this chapter, you analyzed and reflected on how to create opportunities for your team's mathematics assessment evidence to become more transparent and in order to better understand student thinking and mathematics reasoning.

Provided in the next chapter are protocols to help your team develop a similar culture of transparency for mathematics instruction.

CHAPTER 5

How to Lead a Culture of Transparency and Learning With Mathematics Instruction

Improving something as complex and culturally embedded as teaching requires the efforts of all the players, including students, parents, and politicians. But teachers must be the primary driving force behind change. They are best positioned to understand the problems that students face and to generate possible solutions.

—*James W. Stigler and James Hiebert*

In chapter 4, you analyzed and reflected on how to create opportunities for your team's mathematics assessment evidence to become more transparent. In this chapter, you have an opportunity to reflect on protocols that develop a culture of transparency within daily instruction. This type of team transparency impacts the way in which students learn.

According to Harvard researchers, "What predicts performance is what students are actually doing" (City, Elmore, Fiarman, & Teitel, 2009, p. 30). Analyzing student learning requires more than just looking at the end result of the lesson or unit of instruction. Collaborative teams also analyze how students are interacting during lessons. The in-depth analysis of students in action reveals more than what teachers predict students do or do not understand when they are analyzing student work (City et al., 2009).

Your professional leadership responsibility is to intentionally create opportunities and structures for your teams to make their instructional decision making more transparent. Peer observation then becomes the norm in your school culture. Currently, it is rare to observe a school culture where teachers of mathematics consistently observe one another teach the lessons they design together.

Deborah L. Ball, Mark H. Thames, and Geoffrey Phelps (2008) share that teaching is a professional practice that requires knowledge and skill beyond what appears in the essential learning standards for mathematics. Not only do your teams need content knowledge (the most essential learning standards) and the progression of the learning targets in a given unit, but team members also need pedagogical content knowledge on how students make meaning of the content and the instructional strategies that most impact student learning for student engagement, as well as the scope of student needs (Ball, Thames, & Phelps, 2008).

The protocols in this chapter model intentional opportunities to make instruction and instructional practices visible so teams can engage in a continuous cycle of inquiry for instruction.

The research-affirmed elements of high-quality lesson design appear in *Mathematics Instruction and Tasks in a PLC at Work* (Kanold, Kanold-McIntyre, et al., 2018) from this series. In this chapter, you explore two protocols that support building the shared knowledge teacher teams need in order to design effective daily lessons for a positive impact on student learning: (1) instructional rounds and (2) lesson study.

Instructional Rounds

Instructional rounds support the collective team-inquiry process. Typically, teacher observations in the classroom are either informal or formal evaluations by an administrator. Through the evaluation process, the teacher being evaluated is typically the only one expected to learn. With instructional rounds, the expectation is that everyone participating is learning from the experience (City, 2011).

The purpose of instructional rounds is similar to medical rounds that occur during residency. During the instructional rounds, teams of teachers walk from classroom to classroom collecting data on a focused instructional task (for example, questioning, specific instructional strategies, engagement, classroom management, and so on).

Once the rounds are complete, the team of teachers debriefs its observations, provides just-in-time and specific feedback to its peers about the levels of student engagement in the lesson, and formulates action steps based on the evidence of student learning. Instructional rounds can be grade-level based, course based, or cross-curricular depending on the instructional focus. Instructional rounds are beneficial because they (City et al., 2009):

- Break down the isolation of traditional teaching and make teaching a public event

- Require collaborative teams to focus on a common challenge with which each team member wants to improve

- Build capacity with instructional practices and develop a common language related to instruction

- Provide a way to hold teachers accountable for improving instructional practices with feedback in a nonevaluative manner

Instructional rounds promote reflection about the instructional strategies used, and their relationship to student learning actions. The primary purpose and strength of the protocol is the reflection that occurs during the debrief following the observation and the discussions that stem from the observations, and the actions teachers take as they identify next steps (Marzano, 2009). However, without teacher action on the feedback, the instructional round focus will have limited impact on student learning.

As with the other protocols, you will need to establish norms for the instructional rounds. Since the purpose of instructional rounds is to observe student learning, norms ensure observation and feedback tie to evidence of student engagement and learning. The following are some basic norms for instructional rounds.

- When observing in a mathematics classroom, refrain from conversations with the other observers.

- Refrain from assisting students or interrupting the lesson.

- Collect factual observations about student thinking, reasoning and actions that are nonevaluative.

The final point is very important: when you and your team are collecting evidence, your comments and observations during the debrief need to be factual rather than judgmental. For example, you may observe a group of students working on a task and see one of the students is on his phone underneath the desk. A judgmental statement would be, "One student is completely off task and not following the school rules." A more factual statement is something like, "I observed three student teams. One team had one student not engaged in the task." Notice the difference in the tone?

Before you begin using instructional rounds, you and your team need to identify an instructional focus goal. What part of your mathematics instructional vision would you like to explore? What is your school improvement goal? What is the instructional focus from your SMART goal?

For example, one of the research-informed instructional strategies from *Principles to Actions* (NCTM, 2014) is to pose purposeful questions. If you were to observe a mathematics lesson with your team, would each member be in agreement about what posing purposeful questions looks like or sounds like in a classroom?

Start by describing both the teacher and student actions that your team members may observe and what you would like to see during the observation. When you lead the instructional rounds, have your team identify the focus area, come to consensus on what it looks and sounds like during a lesson, and give teachers time to plan lessons that center on the focus goal.

Figure 5.1 is the Instructional Rounds Protocol to complete with your teams as you prepare for the instructional rounds.

The directions for the instructional rounds appear in figure 5.2 (page 92).

Scheduling the instructional rounds and time for your team to debrief is important. If you and your team members are not able to observe one another throughout the day, you may want to observe another team in your building or possibly observe at another school. However you plan it, the observed teachers need to have a clear understanding of the purpose of the instructional rounds and the type of feedback they will receive at the conclusion of the exercise.

Figure 5.3 (page 92) shows an example of combined feedback from the mathematics 1 team to the mathematics 2 team after the mathematics 2 team completes instructional rounds. Both teams are collectively working on engaging students in student-to-student discourse.

Teams or observers:	Date of Instructional Rounds:
Facilitator:	Times:
What are the agreed-on norms?	
What is our instructional focus? • What evidence are we going to collect? • What student actions are being observed that support our team's focused instructional goal? • What teacher actions aligned to the instructional focus promote increased learning or mathematical thinking?	
What is the schedule for the instructional rounds? • Start and end time of observations • Debrief scheduled • Number of classes observed • Number of minutes in each class	
What feedback will be collected and shared?	
Who will collect and share the evidence of the instructional rounds? What expectations for action on the feedback will be promoted?	

Figure 5.1: Instructional Rounds Protocol.

*Visit **go.SolutionTree.com/MathematicsatWork** for a free reproducible version of this figure.*

Protocol Steps (It will take mathematics teams about forty-five minutes for planning, one or two class periods or math lesson time for elementary school for rounds, and forty-five minutes for debriefing.)	Directions
1. Identify the focus area or question the mathematics team is trying to answer, describe evidence the team will be evaluating, and list predictions.	As a mathematics team, choose a focus area—what question are you trying to answer by implementing a new strategy or a new type of task? The following are some sample questions that focus on specific areas. • What student actions support your team's focus (instructional goal)? What actions are not supporting the goal? • What teacher actions promote increased learning or mathematical thinking? • What is the level of engagement? • What mathematical discourse is observable? How effective is the discourse?
2. Collect the evidence.	Agree on the focus (instructional goals) the mathematics team will be observing during the rounds. Create a schedule for the observations. Observations can last from five to twenty-five minutes, depending on the focus goal and the evidence you are seeking to collect. During each observation, collect evidence that aligns to the two to three focus questions.
3. Analyze the evidence.	During the debrief, one person (the facilitator) collects the evidence (electronically, on a whiteboard, or on poster paper). The facilitator encourages mathematics team members to share one positive comment and one comment for consideration (see figure 5.3 for examples), one at a time, until all team members share all of the comments. As a team, look for trends and patterns and discuss next steps based on your observations.
4. Plan action steps based on the results of the analysis.	List two or three action steps based on the discussion during analysis of the evidence.

Figure 5.2: Instructional Rounds Protocol instructions.

*Visit **go.SolutionTree.com/MathematicsatWork** for a free reproducible version of this figure.*

Focus Instructional Goal: How Effective Is Student-to-Student Collaboration During the Mathematics Lesson?

Strengths during the observations:
• There are more students on task (than during the previous instructional rounds).
• There are intentional opportunities for students to engage in partner discussion (for example, partner A tells partner B . . .)
• Teachers move students at the beginning of the lesson to ensure they have peer partners.
• Teachers use *rally coach* structure.
• Teachers use assigned roles during the partner activity to promote equal participation in the activity.
• There are times that teachers pair partners with other pairs to discuss their guesses.
• Teachers use technology such as a document camera to display student work and have students explain their thought processes.
• Teachers address off-task behavior.

Questions to move instructional practices forward:
• How can your team provide wait time so all students have time to think about the questions before sharing with their partner?
• How can you and your team make the thinking visible (students' presentation of their thinking)?
• How can you be intentional about partner selection to ensure that the partners are working effectively?
• Would it be possible to rotate the roles so everyone has an opportunity to work on the computers or summarize the learning?

Figure 5.3: Sample feedback from an instructional round.

Notice that the feedback the mathematics 1 team provides to the mathematics 2 teachers includes statements of strength and questions for things to consider when designing future lessons—questions to move instructional practices forward. This feedback from the mathematics 1 team is nonjudgmental. If you are not observing a full lesson and only see a snapshot of a lesson, feedback should be in the form of questions. Due to the limited lesson observation, the teachers may have addressed the question when you are not in the classroom. The questions should also prompt continued reflection and action for the teachers being observed.

LEADER *Reflection*

What is your teams' mathematics instructional commitment this year? How might you use instructional rounds to support team member implementation of learning about how to implement this instructional expectation?

_____

_____

_____

_____

_____

_____

_____

_____

_____

_____

_____

_____

_____

_____

_____

_____

If your teams engage in multiple instructional rounds during a school year, you can also monitor instructional trends. If you are a coach or site-level leader facilitating the rounds, you can plan additional professional learning based on the trends. Once you and your team feel comfortable with making their instruction public, you and your team are ready to engage in more in-depth analysis of instruction using lesson study.

Lesson Study

Lesson study is a professional learning model that exemplifies the importance of the collaborative process. Lesson studies are an opportunity for teams to collaboratively:

- Co-plan the "best mathematics lesson ever" using ideas from all of the teachers on a team

- Observe the mathematics lesson in action together

- Make revisions to the mathematics lesson based on feedback from your team and evidence of student learning

- Teach the mathematics lesson again to see if the revisions had the predicted impacts

Lesson studies can take many forms, but the premise of a lesson study is not to only focus on improving an individual teacher; rather, it is to collectively improve methods of teaching across a team of teachers. In 1999, James W. Stigler and James Hiebert, in *The Teaching Gap: Best Ideas From the World's Teachers for Improving Education in the Classroom*, identified the seven critical steps of lesson study as follows.

1. **Define the problem:** What do we want to learn about by engaging in the collaborative process?

2. **Plan the lesson:** What resources will we need to focus on our instructional goal?

3. **Teach the lesson:** Which teacher will teach the lesson the first time?

4. **Evaluate the lesson:** What evidence do we observe and collect during the lesson?

5. **Revise the lesson:** What might we change in the lesson to improve student understanding?

6. **Teach the revised lesson:** Who will teach the lesson the second time?

7. **Evaluate and reflect again:** What might we still change in the lesson to improve student learning?

Lesson study differs from instructional rounds because during lesson study, you and your team are observing a team-created lesson; during instructional rounds, the lesson can differ, but the instructional focus is common.

Lesson-Plan Design and Analysis

The first step of lesson study is to choose an instructional goal and define what content or instructional strategy you would like to learn more about. Using the Mathematics in a PLC at Work lesson-design tool available online (visit **go.SolutionTree.com/Mathematics atWork**), work with your team to collaboratively plan a lesson that aligns to the instructional or content focus (the second step of the process). (The lesson-design tool also appears in *Mathematics Instruction and Tasks in a PLC at Work* in the *Every Student Can Learn Mathematics* series; see Kanold, Kanold-McIntyre, et al., 2018.) Use the Team Discussion Tool: Lesson-Study Protocol (figure 5.4) to assist you with planning the lesson.

Lesson-Study Protocol		
Read through the lesson plan and look for evidence to support the specific components of a well- planned lesson. Provide feedback to your colleagues to make the best lesson possible.		
Lesson Element	**Probing Questions**	**Comments, Questions, or Clarifications**
Preparing for the **Why** of the Mathematics Lesson	• What are the essential mathematics learning standards and learning target or targets for the lesson? (What do you want students to know and understand about mathematics as a result of this lesson?) • What is the process standard of the lesson? How will students engage in the process?	
Beginning-of-Mathematics-Class Routines	• In what ways does the task build on students' prior knowledge, life experiences, and culture? • What are the definitions, concepts, or ideas students need to know to begin to work on the task? How will you develop students' academic language and vocabulary? • What questions will you ask to help students access their prior knowledge and relevant life and cultural experiences? • What are the connections between important ideas in this lesson and important ideas in past and future lessons?	
Instruction: During-Class Routines	• What are the specific activities, investigations, problems, questions, or tasks students will work on during the lesson? • How are the mathematical procedures in the lesson justified and connected with important ideas? • What are all the ways students can solve the task? • Which solution pathways will students use? • What are potential student misconceptions? • What are potential student errors? • What are assessing questions or prompts you can use to get students "unstuck"? • What are advancing questions or prompts to further student understanding? • How will you engage all students to think about your questions? • How do you see or hear students engage with mathematical ideas during class? • What is the expectation of student engagement for sense-making and reasoning? • How will you ensure the task is accessible to all students while still maintaining a high cognitive demand for students? • What will be the student-to-student interaction you employ? • How does the plan address the class discussion? • Which solution pathways do you want to share during the class discussion? In what order will you present the solutions? Why? • What opportunities exist for students to productively struggle with the mathematical ideas? • How do you respond to students' struggles, and how does your response support students to stay engaged in the mathematical tasks for the lesson?	

| End-of-Class Routines | • What are the student-led summary activities, questions, and discussions to close the lesson and provide a foreshadowing of future learning?
• How will students self-assess their understanding?
• How do students know that they met the learning target for the day?
• What formative assessment strategies will you employ during the lesson?
• What evidence will you use to determine the level of student learning of the daily learning target? | |

Figure 5.4: Team discussion tool—Lesson-Study Protocol.

Visit **go.SolutionTree.com/MathematicsatWork** *for a free reproducible version of this figure.*

Once you and your team plan the lesson with an instructional goal and set the date for the lesson study, you are ready to prepare for facilitation of the lesson study.

Lesson-Study Planning

If engaging in lesson study is new for you and your mathematics team, identifying a strong facilitator is necessary. Your lesson-study facilitator needs to be knowledgeable and tactful and someone who can manage conversations with a respectful tone. The facilitator will model an analytical approach during the observations and debriefs of the lesson.

It may be appropriate to have an outside expert, a knowledgeable peer who is not on the team, to facilitate the first lesson study. If you are a coach or site- or district-level leader, you can facilitate the first few lesson studies to model how to adhere to the norms and process for reflection and revision.

When you and your team become familiar with the protocol, peers can effectively lead lesson studies with peer-to-peer team interaction if you are a team leader new to the Lesson-Study Protocol.

Before you begin instruction of the co-planned lesson, clearly articulate team norms to honor the work of the demonstration teacher and the efforts of the team members who create the lesson.

Consider the sample norms for the observation and debrief of the lesson study in figure 5.5 (page 96). Which norms would be beneficial for your team? What else might you add to the list?

Norms for teacher interactions during the Lesson-Study Protocol are vital for ensuring the observations and comments during the debrief session further team learning. During the observation of the lesson, there are two crucial norms to note for the observers.

Personal Story **JESSICA KANOLD-McINTYRE**

The first time I facilitated a lesson-study protocol, I learned quite a bit about what *not* to do. I was working with a strong teacher team I had a positive relationship with, and I knew team members trusted each other. So I briefly passed over the norms and jumped right into the process of planning a lesson in order to give them time to work.

What I had failed to realize was how important the norming conversation would be to help set the stage for the actual day of observing. On the observation day, I quickly learned that the two norms of not interacting with students during the observation and preparing the team for how to provide feedback on the lesson are crucial to the success of the lesson study.

The demonstration teacher was extremely nervous about the observation because she was worried what her colleagues would say about her teaching. If I would have taken time for the team to process how the lesson study would progress, team members would have set their norms for how to respond and provide feedback about the lesson. This demonstration teacher would have had much more confidence during the observation.

Lesson-Study Component	Suggested Norms
Evaluating the Mathematics Lesson	Observers should not interfere with the natural process of the lesson (for example, by helping students with a problem).
	Observers may circulate around the classroom when students are engaged in an extended task in their group. (The purpose is to listen to what students are saying and watch what they are doing and to document the evidence of learning or challenges students are facing.) Otherwise, observers should stand to the back and sides of the classroom.
	The demonstrating teacher should provide a seating chart or name tags for students, which is helpful for observers.
	Observers refrain from talking or carrying on sidebar conversations during the lesson. Each observer gets a small group of students to monitor and capture student thinking.
Revising the Mathematics Lesson	Choose a timekeeper, recorder, and other roles as you need to.
	Each observer may address one element of the lesson within a predetermined time limit. During his or her time to speak, others will listen and take notes (no cross talk). Provide constructive feedback—be specific and nonjudgmental.

Figure 5.5: Sample norms for lesson study.

First, the most important norm is that you and your team members evaluating the lesson *do not* interact with the students. This is probably the most challenging norm! A lesson study is action research; when you interrupt students and the learning process, you will not know if the planned lesson elements impacted the students or if the observers around the room made the difference.

Second, since the lesson study is about the impact of the lesson on student learning, encourage your mathematics team members to focus comments on the lesson versus the teacher. For example, instead of starting a comment with, "You did . . ." or "Your lesson . . .", use comments such as:

- "This element was a strength of the lesson because . . ."

- "I noticed . . ."

- "I wonder if . . ."

The purpose of the lesson study is to learn how each element of the planned lesson impacts student learning. Positive team interactions are vital to your team's growth.

Before teaching the lesson, review the Lesson-Study Protocol directions in figure 5.6 and plan accordingly. As you read through the instructions, consider the planning you need to undertake to complete the protocol.

Figure 5.7 (page 98), the lesson-study student evidence form, is the data-collection tool for observing demonstration lessons. This form closely mirrors the Mathematics in a PLC at Work lesson-design tool (available at **go.SolutionTree.com/MathematicsatWork**) that you and your team used to design the lesson. This is to help you take notes in a way that mirrors the flow of the lesson you planned. Review figure 5.7 as a team, and discuss what specific evidence you and your team anticipate observing.

Consider the leader reflection to help you prioritize.

LEADER *Reflection*

What types of special scheduling will be needed to help my teams plan and observe both the demonstration lesson and then the follow-up lessons? How will I provide time and facilitate teacher planning and the debrief of both lessons?

_____

_____

_____

_____

_____

_____

_____

_____

_____

Protocol Steps (Time varies depending on the amount of time for planning, teaching the lesson, debriefing, and teaching the revised lesson.)	Directions
1. Identify the focus area or the question the team is trying to answer, describe evidence team members will evaluate, and list predictions.	**Plan the Lesson** Use this protocol to answer the following questions. • What is the instructional focus for your lesson study? • What content or process standards do you want to research to improve student understanding? • Which instructional strategies will you need to employ to ensure equal participation from each student? • What tasks will elicit evidence of student understanding that align to the essential learning standard?
2. Collect the evidence.	**Evaluate the Lesson** Do the following. • Print copies of the co-planned lesson and the evidence form (figure 5.7, page 98) for each observer. • Set up chairs around the room for each observer. • Assign each observer to a set of students to observe and record information related to learning to ensure each student is being observed. • Collect evidence of student thinking and engagement in each element of the lesson. • As an observer, do not interfere with the lesson or assist students when observing.
3. Analyze the evidence.	**Debrief the Observations From the Lesson** For the debrief: • Reflect on each element of the lesson (demonstration teacher and observers). • Comment on student data observations (observers) and share comments at the end of each round (the demonstration teacher). The facilitator summarizes the comments and possible revisions at the end of each element of the lesson plan. • Record comments and possible revisions to the lesson plan. • Continue the cycle of reflect, refine, and act until the group has discussed the last element of the lesson. • Clarify student evidence that team members must collect to determine if the modifications teams make will affect student learning during the revised lesson. • Revise the lesson and assign another teacher to teach the revised lesson. • Engage in a second debrief of the revised lesson following the same protocol as the first debrief.
4. Plan action steps based on the analysis results.	**Next Steps** • After the debrief of the original and modified lessons, complete an individual reflection and team reflection related to lesson study using figure 5.9 (page 100). • Identify specific teacher actions tied to the instructional focus that each team member can use to determine how to commit to continued improvement and provide clarity for next steps.

Figure 5.6: Lesson-Study Protocol instructions.

*Visit **go.SolutionTree.com/MathematicsatWork** for a free reproducible version of this figure.*

Lesson-Study Student Evidence Form

Directions: Collaboratively complete the Mathematics in a PLC at Work lesson-design tool (available at **go.SolutionTree.com/MathematicsatWork**) and provide a copy to each observer. During the demonstration lesson, collect evidence of student learning, discussions, and student misconceptions.

Beginning-of-Class Routines: What Evidence of Student Thinking and Engagement Have You Observed?

Learning target: How do the students consider the why of the lesson and identify the learning target during the lesson? How do students engage in the prior-knowledge task?

Academic language and vocabulary: Describe the academic language and vocabulary students will use and be taught.

Instruction: During-Class Routines—What Evidence of Student Thinking and Engagement Have You Observed?

Task 1: Cognitive Demand (Circle one) **High** or **Low**

What are the learning activities to engage students in learning the target? Be sure to list materials you need, if necessary.

Student Actions How are the students actively engaged in each part of the lesson? What type of student discourse structure do you observe?	**Questioning** What are the assessing and advancing questions you observed for each task?	**Assessment** What feedback do you observe? How does the feedback move thinking forward and keep the students engaged?

Task 2: Cognitive Demand (Circle one) **High** or **Low**

What are the learning activities to engage students in learning the target? Be sure to list materials you need, if necessary.

Student Actions How are the students actively engaged in each part of the lesson? What type of student discourse structure do you observe?	**Questioning** What are the assessing and advancing questions you observed for each task?	**Assessment** What feedback do you observe? How does the feedback move thinking forward and keep the students engaged?

Task 3: Cognitive Demand (Circle one) **High** or **Low**

What are the learning activities to engage students in learning the target? Be sure to list materials you need, if necessary.

Student Actions How are the students actively engaged in each part of the lesson? What type of student discourse structure do you observe?	**Questioning** What are the assessing and advancing questions you observed for each task?	**Assessment** What feedback do you observe? How does the feedback move thinking forward and keep the students engaged?

End-of-Class Routines: What Evidence of Student Thinking and Engagement Was Observed?

Lesson closure for evidence of learning: Based on the student-led closure, did the students demonstrate proficiency for the daily learning target, and how do you know?

Figure 5.7: Lesson-study student evidence form.

*Visit **go.SolutionTree.com/MathematicsatWork** for a free reproducible version of this figure.*

Lesson study is the ultimate *reflect, refine, and then act* teacher learning process. However, please consider the following about the time demands of lesson study. The first time a team comes together to co-plan a lesson, it will likely take more time to get every team member's perspective into one lesson. Encourage the team to work on the lesson two to four weeks *prior* to the scheduled day of observing the lesson in action. Subsequent lesson studies generally do not require as much planning time. Figure 5.8 (page 100) includes an example of the time commitments the team needs for each element of a lesson study.

What is the team's takeaway from facilitating or participating in the lesson study? How did observing the lessons challenge each team members' assumptions about teaching and learning? After your teams engage in the lesson study, it is important to reflect and receive feedback from the debrief sessions and to consider how you are moving closer to your instructional commitment for the school year.

Figure 5.9 (page 100) is a lesson-study reflection activity each team member can complete individually at the end of the debrief session. Once completed, each participant can share his or her learning and brainstorm ideas about next steps for the team.

As you lead your teams in making instruction more transparent, instructional rounds and lesson studies are two formal protocols for fine-tuning aspects of instruction.

Due to the time demands of lesson studies (see figure 5.8, page 100), you may need to vary the protocol to meet your team and site needs. The following are a few examples of how to vary the protocol.

- The mathematics team chooses a task and co-plans a lesson; however, each team member teaches the lesson independently. Once all team members teach the lesson, they each bring the student work and any individual student or group learning observations back to the team for discussion.

- One team member records himself or herself teaching the lesson. One note of caution: Recording the lesson is not as beneficial as a live lesson because of the limitation of capturing *all* student thinking and reasoning during the lesson. You might consider asking someone on campus to take a video of the lesson for the team so it can record multiple student groups as they are discussing solutions and also capture student work in action.

Both approaches still ensure teams are making instructional conversations a priority and are viable strategies for examining instructional and assessment practices based on what the group has learned from looking at student work.

In addition to instructional rounds and lesson study, you can consider the following more informal and less intense strategies to make instruction more transparent.

- **Complete peer observations:** Create a schedule for everyone on the team to observe the other teachers on their team.

- **Observe another team:** Work with team leaders, coaches, or site-level leaders to arrange observing another team on campus or at another school.

- **Focus instructional practice:** Choose a focus instructional practice; and during the week, have each team member go observe two other teachers and then debrief the process at the next team meeting.

- **Coteach a class first period:** Have each teacher on the team coteach the first period of the school day. Then, each teacher will teach the same lesson separately throughout the rest of the day.

- **Participate in the #ObserveMe challenge:** Mathematics teacher specialist and trainer Robert Kaplinsky (2016) started the #ObserveMe challenge for those on Twitter. To participate, team members post feedback requests outside their classroom door, take a picture of it, and tweet it using the hashtag #ObserveMe. Kaplinsky (2016), for example, posted "How can I improve the way I set up a problem to allow students to become engaged without immediately becoming overwhelmed?" Other teachers can stop by any time to provide feedback related to the question. For more information and additional examples of questions, search #ObserveMe on Twitter.

Lesson-Study Element	Amount of Time Needed	Expectation
Defining the problem	One hour	Determine what we want to learn by engaging in the collaborative process.
Planning the lesson	Two to four weeks	Plan the lesson prior to the scheduled lesson-study day; include the resources we will need for the lessons.
Previewing the lesson	One hour	Preview the lesson so every observer understands what to expect in the lesson; review norms.
Teaching the lesson	One class period	Decide which teacher will teach the lesson the first time for evaluation of effectiveness. That person then teaches the lesson.
Analyzing the evidence: 1. Evaluating the lesson 2. Revising the lesson	Two to three hours	Debrief the first lesson and make revisions to the lesson plan. Evaluate the evidence we observe and collect during the lesson. Understand what we might change in the lesson to improve student understanding.
Teaching the revised lesson	One class period	Determine who will teach the lesson the second time. That person then teaches the revised lesson.
Evaluating and reflecting again	One hour	Debrief the lesson study and plan action steps.

Figure 5.8: Time expectations of a lesson study.

Lesson-Study Team Reflection

Each team member should respond to the following questions individually at the end of the debrief session. Once completed, each team member can share his or her responses about next steps for the team.

1. **What:** What did you learn from the mathematics lesson study?

2. **So what:** How did the mathematics lesson study challenge your assumptions or expectations?

3. **Now what:** Where do we go from here? What's the next step of our daily mathematics lesson designs?

Figure 5.9: Lesson-study team reflection.

*Visit **go.SolutionTree.com/MathematicsatWork** for a free reproducible version of this figure.*

Success in Making Learning Public

The protocols featured in this chapter support your mathematics team engagement in reflective practices around core mathematics instruction. When making instruction more transparent, your team begins to develop shared understanding of the depth of the essential learning standards and connect the learning targets to the instructional experiences of the students.

Additionally, utilizing instructional rounds or lesson studies provides leaders with the opportunity to support the team's learning needs and collectively respond to the third and fourth critical questions of a PLC (DuFour et al., 2016) in relation to the team or team members you serve.

3.   How will we respond when some *team members* do not learn?

4.   How will we extend the learning for *team members* who are already proficient?

Use the leader reflection to consider actions you can take if teams continue to struggle with instructional practices.

LEADER *Reflection*

If teams are still struggling to share instructional practices and become more transparent with one another about their daily mathematics lessons, what actions will you take as the team leader, coach, or site-level leader? If your teams are excelling at transparency in their practice, what actions will you take to make their work more public, and use their work as exemplars for others?

_____

_____

_____

_____

_____

_____

_____

_____

TEAM RECOMMENDATIONS

Leading in a Culture of Instructional Transparency

- Choose protocols for team engagement based on your team's readiness level to become more transparent with one another about their mathematics instruction.

- Use the protocol two or three times before moving to a new protocol so mathematics team members can engage in more in-depth conversations about student learning.

- Ensure the mathematics team members have time for reflection based on observed evidence of student learning.

- Work together with other leaders to monitor and support team reflection on observed evidence of student learning and require team action based on evidence.

As you help your teachers to reflect, refine, and act with instructional rounds and lesson study, they will discover they are not alone in trying to solve the complex and daily problems they face, and they will learn to become more vulnerable and open with one another. They also will learn to celebrate each other's successes along the way. The small victories become more public and less private. And, the quality of student learning for mathematics begins to rise.

PART 2 SUMMARY

Collaboration is a continuous, job-embedded, ongoing process for the mathematics learning culture *you lead*. Your collaborative teams' efforts are for the purpose of developing team member knowledge to support guaranteed equity in student access to quality instruction, meaningful feedback through intentional lesson design, and coherent learning outcomes. Your ultimate responsibility is to create a culture that will sustain success for each and every mathematics learner while simultaneously cultivating a positive mathematical identity and high sense of agency for each and every student.

In part 1 of this book, you explored personal leadership *practices* you develop in order to provide supportive conditions for effective collaboration. Your role includes creating a plan to address the critical foundation and measures of success you will continuously monitor to ensure student learning.

As you develop your individual leadership practices, you also ensure the collaborative teams focus on the right work using effective leadership *strategies*. Effective strategies promote strong mathematics team communication and establish a clear purpose for your team's artifacts, collaboration, and continued professional learning.

In part 2 of this book, you addressed protocols necessary to embrace a culture of transparency. Each protocol aimed to support your team's growth in instructional and assessment practices as team members learn from one another through the lens of analyzing student work and evidence of learning.

When your teams are engaged in deep reflection, team members will share evidence of learning from unit assessment data or teach lessons as part of learning from one another as a learning laboratory for the profession, to refine their own understanding of how students learn mathematics.

Epilogue

By Timothy D. Kanold

An epilogue should serve as a conclusion to what has happened to you based on your experiences in reading and using this book. An epilogue should also serve as a conclusion to your progress and change toward leading the growth of your mathematics programs and your teachers of mathematics each and every school season as your professional leadership and career unfold.

During the 1990s and early 2000s, my colleagues and I at Adlai E. Stevenson High School District 125—the birthplace of the PLC at Work model—developed the deep mathematics experiences and the levels of transparency required that are revealed in this book. Stevenson eventually became a U.S. model for other schools and districts because of the deep inspection, revision, and eventual collective team response to the quality of the common unit-by-unit assessments, interventions, and daily lessons designed in mathematics for each grade level and course, K–12.

At the time, many considered our decision to become more transparent in our lesson-design practice, go into each other's classrooms, and to analyze student performance on unit assessments together as a groundbreaking practice. To be sure, we were quite hesitant. What would we discover? What if we could not measure up to the work ethic of our other colleagues?

What we discovered, however, is that we were all in the same professional boat, struggling with the same frustrations. More importantly, we discovered that when we *collectively* pursued our mathematics teaching and assessing practices with certain ferocity, student performance began to soar. Soar! There is nothing more rewarding than the thrill of improved student learning due to your effort and action. The talented coauthors of this book and of our series will tell you they experienced the same results in their schools and districts.

Developing the leadership to create a more transparent practice built on the four critical questions of a PLC culture, using high-quality mathematics assessments with just-in-time student interventions, and designing daily lessons together for learning each standard became part of our leadership responsibilities and our professional practice.

But these practices did not exist for us initially. And, as the director of mathematics, I did not really know how to lead the effort forward either. I was new to the idea of becoming a school mathematics leader. I just did not know where to start.

This book, in conjunction with the other three books of our *Every Student Can Learn Mathematics* series, can save you some time as it provides a roadmap for your own leadership growth, and a framework for action that guarantees to significantly improve student learning in mathematics overnight. But, you will need to lead the effort, and your mathematics teachers will need to be pushed and pulled toward the "right things" we promote in the series.

At the time, I could not have articulated it this clearly, but there were two areas of improvement that needed to change in our schools and district. The first was helping every teacher of mathematics become more reflective as professionals and as practitioners. We did not know how to *reflect, refine, and act* on the quality of our work. We just acted all of the time, with not much reflection.

You can change that culture in your school. Just use the advice within the covers of this book.

Second, we did not know where to start; there were just so many holes in the dam, so to speak. *So here is my recipe for mathematics program improvement, in this order.*

These first three parts of the recipe are provided in our *Mathematics Assessment and Intervention in a PLC at Work* book of this series.

1. Examine every mathematics unit exam given to students and improve the quality and the rigor of those local assessments.

2. Score all local assessments with at least one other teacher of mathematics to check for accuracy in calibration of scoring.

3. Foster student ownership and agency in all assessment results through on ongoing analysis of unit-by-unit mathematics standard performance.

These next three parts of the mathematics program recipe are provided in our *Mathematics Instruction and Tasks in a PLC at Work* book of this series.

4. Decide to never go back and teach mathematics lessons using bad practice and routines.

5. Understand that research informs five or six essential elements of every lesson and makes those elements non-negotiable, but with lots of freedom for implementation. As a leader you cannot accept less than best practice in mathematics instruction from every teacher.

6. Share those best lesson-design elements among all teachers and engage in formative feedback discussions about those essential lesson-design elements for student perseverance during the lesson.

These next three parts of the mathematics program recipe appear in the *Mathematics Homework and Grading in a PLC at Work* book of this series.

7. Decide to do the transparent work of discussing the homework and grading in mathematics.

8. Like mathematics assessments, make homework a common artifact for every teacher team. The common unit-by-unit homework assigned is based on research and understanding of formative independent practice.

9. Revise homework, make-up work, and grading routines to inspire and not destroy student learning.

And, finally, the last part includes coaching and collaboration.

10. Mix it all together with dynamic mathematics leaders—that means you!

(Visit **go.SolutionTree.com/MathematicsatWork** to download a free copy of this recipe: the Mathematics in a PLC at Work Recipe for Improved Student Learning.)

Ultimately, you will need to decide how to use this recipe. My advice is to think short and long term simultaneously. Of all the parts of this recipe, what is most urgent and needs attention now, and what needs to be part of a three-year implementation plan (or more)? Celebrate your progress as you reflect. Ask, How can we refine and improve our current efforts?, and then take action once again. The very nature of our work is cyclical. Our leadership journey is to always make that next cycle just a little bit better. We hope this book and the series help you along a journey that pursues *Every Student Can Learn Mathematics* as its fundamental purpose and goal.

On behalf of Mona, Matt, Bill, Jessica, Sarah, and myself, may your mathematics leadership journey lead to inspired student learning each and every day.

Cognitive-Demand-Level Task Analysis Guide

Table A.1: Cognitive-Demand Levels of Mathematical Tasks

Lower-Level Cognitive Demand	Higher-Level Cognitive Demand
Memorization Tasks • These tasks involve reproducing previously learned facts, rules, formulae, or definitions to memory. • They cannot be solved using procedures because a procedure does not exist or because the time frame in which the task is being completed is too short to use the procedure. • They are not ambiguous; such tasks involve exact reproduction of previously seen material and what is to be reproduced is clearly and directly stated. • They have no connection to the concepts or meaning that underlie the facts, rules, formulae, or definitions being learned or reproduced.	**Procedures With Connections Tasks** • These procedures focus students' attention on the use of procedures for the purpose of developing deeper levels of understanding of mathematical concepts and ideas. • They suggest pathways to follow (explicitly or implicitly) that are broad general procedures that have close connections to underlying conceptual ideas as opposed to narrow algorithms that are opaque with respect to underlying concepts. • They usually are represented in multiple ways (for example, visual diagrams, manipulatives, symbols, or problem situations). They require some degree of cognitive effort. Although general procedures may be followed, they cannot be followed mindlessly. Students need to engage with the conceptual ideas that underlie the procedures in order to successfully complete the task and develop understanding.
Procedures Without Connections Tasks • These procedures are algorithmic. Use of the procedure is either specifically called for, or its use is evident based on prior instruction, experience, or placement of the task. • They require limited cognitive demand for successful completion. There is little ambiguity about what needs to be done and how to do it. • They have no connection to the concepts or meaning that underlie the procedure being used. • They are focused on producing correct answers rather than developing mathematical understanding. • They require no explanations or have explanations that focus solely on describing the procedure used.	**Doing Mathematics Tasks** • Doing mathematics tasks requires complex and no algorithmic thinking (for example, the task, instructions, or examples do not explicitly suggest a predictable, well-rehearsed approach or pathway). • It requires students to explore and understand the nature of mathematical concepts, processes, or relationships. • It demands self-monitoring or self-regulation of one's own cognitive processes. • It requires students to access relevant knowledge and experiences and make appropriate use of them in working through the task. • It requires students to analyze the task and actively examine task constraints that may limit possible solution strategies and solutions. • It requires considerable cognitive effort and may involve some level of anxiety for the student due to the unpredictable nature of the required solution process.

Source: Smith & Stein, 1998. Used with permission.

Mathematics in a PLC at Work Rating and Reflection Framework

Mathematics Team Actions Serving the Four Critical Questions of a PLC: Evidence of the Reflect, Refine and Act Process at Team Meetings and in Classrooms

Grade-level or course-based team: _____   Team members: _____   Date: _____

	Equitable Assessment and Intervention		Equitable Instruction and Tasks	
	Team action 1: Develop high-quality common assessments for the agreed-on essential learning standards.	**Team action 2:** Use common assessments for formative student learning and intervention.	**Team action 3:** Develop high-quality mathematics lessons for daily instruction.	**Team action 4:** Use effective lesson designs to provide formative feedback and student perseverance.
Team Meeting With Documents	☐ Develop common balanced assessments aligned to and organized by the essential standards for the unit. ☐ Develop assessments that have a balance of tasks, item types, and align to both the content and process demands of the standards. ☐ Create common assessments with common scoring agreements and set proficiency for each essential standard.	☐ Discuss feedback to students and the impact of that feedback. ☐ Develop a protocol or template for students to fix or embrace their errors from the common assessments and identify what they have learned or not learned yet to set goals and take action. ☐ Create a plan to re-engage students with essential learning standards that they may not have mastered.	☐ Choose common higher- and lower-level-cognitive-demand tasks that align to the essential standards of the unit. ☐ Determine effective instructional strategies to use during lessons. ☐ Align tasks and instructional strategies for learning the standards in the unit. ☐ Identify the mathematical language and thinking for how students will engage in the learning targets. ☐ Discuss tasks and rationale for the six components of quality lesson design.	☐ Work mathematics tasks together and identify scaffolding and advancing prompts for the tasks. ☐ Discuss intentional differentiated and targeted in-class Tier 1 supports as students engage in higher-level-cognitive-demand tasks. ☐ Plan to develop student reasoning and sense making through the formative assessment process. ☐ Plan for teacher feedback with student action using small-group discourse during lessons.
	1 2 3 4	1 2 3 4	1 2 3 4	1 2 3 4
Across Team Classrooms	☐ Students reference the same learning target as teachers. ☐ Students can articulate the learning target to each other. ☐ Students take the same assessment on the same day. ☐ Students can explain the learning standards and the expectations for meeting the proficiency on each target.	☐ Students complete their reflection and goal tracker using the learning targets, common assessment data, and formative feedback. ☐ Students create a learning plan in each classroom and take action on that plan. ☐ Students can describe the re-engagement plan.	☐ Same general pacing of the lessons is evident, and teachers use similar higher-level-cognitive-demand tasks. ☐ Teachers use consistent instructional strategies in lessons across all classrooms. ☐ Students engage in similar process standards through the use of a task and are using academic language.	☐ Time is built into lessons for small-group discourse that provides student-to-student feedback and teacher-to-student feedback with student action. ☐ Students receive consistent differentiated prompts during their work on a higher-level-cognitive-demand task.
	1 2 3 4	1 2 3 4	1 2 3 4	1 2 3 4
Rating				

Recommended next steps for your team as you address the assessment and instruction-related team actions to improve student learning:

1 = Beginning; 2 = Practicing; 3 = Implementing; 4 = Embracing

	Equitable Homework and Grading		Leading Coaching and Collaboration	
	Team action 5: Develop and use high-quality common independent practice assignments for formative student learning.	**Team action 6:** Develop and use high-quality common grading components and formative grading routines.	**Coaching action 1:** Develop PLC structures for effective teacher team engagement, transparency, and action.	**Coaching action 2:** Use common assessments and lesson-design elements for teacher team reflection, data analysis, and subsequent action.
Team Meeting With Documents	□ Collaboratively plan homework assignments for a unit before the unit begins and establish purpose. □ Use an agreed-on homework protocol to determine a limited number of tasks, spaced practice, balance of cognitive demand, and alignment to essential standards. □ Develop common scoring agreements on homework. 1 2 3 4	□ Use an agreed-on system for determining student grades. Develop common grading practices and design effective grading practices. □ Come to agreement on the purpose of a grade. □ Establish grading systems to provide meaningful feedback to all stakeholders. 1 2 3 4	□ Collaboratively set SMART goals and norms. □ Articulate clear expectations for team actions at the beginning of the school year (for example, using the team-building worksheet). □ As a team, discuss the six team actions to collectively answer the four critical questions of a PLC. 1 2 3 4	□ Determine which students did and did not meet proficiency targets for essential learning standards. □ Collectively respond to the analyzed data with a specific plan for continued student learning during the next unit. □ Engage in reflection on impact of instruction using an established protocol. 1 2 3 4
Across Team Classrooms	□ Team members assign homework in each classroom (provided to students with answers in advance of starting the unit). □ Students work together across classes on homework. □ Students use homework as a tool for independent formative feedback and practice. 1 2 3 4	□ Students discuss grades in terms of student learning rather than point collection. Teachers clearly articulate evidence of student learning to students. □ Students use the feedback from homework and grades to monitor their learning. □ Scoring in one classroom is equivalent to scoring in another. 1 2 3 4	□ Teachers use similar instructional strategies to meet SMART goals. □ Teachers use similar artifacts that support team learning. 1 2 3 4	□ Students are re-engaged in Tier 2 interventions to address learning needs by standard. □ Students similarly re-engage in learning activities in each classroom (Tier 1 interventions and enrichments). 1 2 3 4
Rating	1 2 3 4	1 2 3 4	1 2 3 4	1 2 3 4

Recommended next steps for your team as you address the homework, grading, and coaching team actions to improve student learning:

1 = Beginning; 2 = Practicing; 3 = Implementing; 4 = Embracing

Source: Adapted from Kanold, 2014.

Figure B.1: Mathematics in a PLC at Work rating and reflection framework.

Visit go.SolutionTree.com/MathematicsatWork for a free reproducible version of this figure.

References and Resources

Aguilar, E. (2013). *The art of coaching: Effective strategies for school transformation.* San Francisco: Jossey-Bass.

Ball, D. L., Thames, M. H., & Phelps, G. (2008). Content knowledge for teaching: What makes it special? *Journal of Teacher Education, 59*(5), 389–407.

Barnes, B., & Toncheff, M. (2016). *Activating the vision: The four keys of mathematics leadership.* Bloomington, IN: Solution Tree Press.

Barth, R. S. (2001). *Learning by heart.* San Francisco: Jossey-Bass.

City, E. A. (2011). Learning from instructional rounds. *Educational Leadership, 69*(2), 36–41. Accessed at www.ascd.org /publications/educational-leadership/oct11/vol69/num02/Learning-from-Instructional-Rounds.aspx on June 25, 2017.

City, E. A., Elmore, R. F., Fiarman, S. E., & Teitel, L. (2009). *Instructional rounds in education: A network approach to improving teaching and learning.* Cambridge, MA: Harvard Education Press.

Clifford, C. (2017, May 18). *Why Wharton's no. 1 professor recommends keeping a resume of your failures.* Accessed at https:// cnbc.com/2017/05/18/wharton-professor-adam-grant-says-keep-a-resume-of-failures.html on September 5, 2017.

Conzemius, A. E., & O'Neill, J. (2014). *The handbook for SMART school teams: Revitalizing best practices for collaboration* (2nd ed.). Bloomington, IN: Solution Tree Press.

Duckworth, A. (2016). *Grit: The power of passion and perseverance.* New York: Scribner.

DuFour, R. (2015). *In praise of American educators: And how they can become even better.* Bloomington, IN: Solution Tree Press.

DuFour, R., DuFour, R., & Eaker, R. (2006). *Professional Learning Communities at Work plan book.* Bloomington, IN: Solution Tree Press.

DuFour, R., DuFour, R., Eaker, R., & Many, T. (2006). *Learning by doing: A handbook for Professional Learning Communities at Work.* Bloomington, IN: Solution Tree Press.

DuFour, R., DuFour, R., Eaker, R., & Many, T. (2010). *Learning by doing: A handbook for Professional Learning Communities at Work* (2nd ed.). Bloomington, IN: Solution Tree Press.

DuFour, R., DuFour, R., Eaker, R., Many, T. W., & Mattos, M. (2016). *Learning by doing: A handbook for Professional Learning Communities at Work* (3rd ed.). Bloomington, IN: Solution Tree Press.

DuFour, R., & Eaker, R. (1998). *Professional Learning Communities at Work: Best practices for enhancing student achievement.* Bloomington, IN: Solution Tree Press.

DuFour, R., & Marzano, R. J. (2011). *Leaders of learning: How district, school, and classroom leaders improve student achievement.* Bloomington, IN: Solution Tree Press.

Duhigg, C. (2016, February 25). What Google learned from its quest to build the perfect team. *The New York Times Magazine.* Accessed at https://nytimes.com/2016/02/28/magazine/what-google-learned-from-its-quest-to-build-the -perfect-team.html?mwrsm=amp-email&_r=0 on September 12, 2017.

Erkens, C., & Twadell, E. (2012). *Leading by design: An action framework for PLC at Work leaders.* Bloomington, IN: Solution Tree Press.

Feiman-Nemser, S. (1983). Learning to teach. In L. Shulman & G. Sykes (Eds.), *Handbook of teaching and policy* (pp. 150–170). New York: Longman.

Feiman-Nemser, S. (2012). *Teachers as learners*. Cambridge, MA: Harvard Education Press.

Frick, D. M., & Spears, L. C. (Eds.). (1996). *On becoming a servant-leader: The private writings of Robert K. Greenleaf*. San Francisco: Jossey-Bass.

Fullan, M. (2001). *Leading in a culture of change*. San Francisco: Jossey-Bass.

Gallimore, R., & Ermeling, B. A. (2010, April 14). Five keys to effective teacher learning teams. *Education Week*. Accessed at www.edweek.org/ew/articles/2010/04/13/29gallimore.h29.html/html on June 20, 2017.

Given, H., Kuh, L., LeeKeenan, D., Mardell, B., Redditt, S., & Twombly, S. (2009). Changing school culture: Using documentation to support collaborative inquiry. *Theory Into Practice, 49*(1), 36–46.

Goleman, D. (2011). *Leadership: The power of emotional intelligence—Selected writings*. Northampton, MA: More Than Sound.

Graham, P., & Ferriter, B. (2008). One step at a time: Many professional learning teams pass through these 7 stages. *Journal of Staff Development, 29*(3), 38–42.

Grit. (n.d.). In *Merriam-Webster's online dictionary*. Accessed at https://merriam-webster.com/dictionary/grit on June 8, 2017.

Grover, R. (Ed.). (1996). *Collaboration: Lessons learned series*. Chicago: American Library Association.

Gutiérrez, R. (2002). Enabling the practice of mathematics teachers in context: Toward a new equity research agenda. *Mathematical Thinking and Learning, 4*(2–3), 145–187.

Hansen, A. (2015). *How to develop PLCs for singletons and small schools*. Bloomington, IN: Solution Tree Press.

Hattie, J., (2012). *Visible learning for teachers: Maximizing impact on learning*. New York: Routledge.

Heath, C., & Heath, D. (2010). *Switch: How to change things when change is hard*. New York: Broadway Books.

Hirsh, S., & Killion, J. (2007). *The learning educator: A new era in professional learning*. Oxford, OH: Learning Forward.

Kanold, T. D. (2011). *The five disciplines of PLC leaders*. Bloomington, IN: Solution Tree Press.

Kanold, T. D. (2017). *HEART! Fully forming your professional life as a teacher and leader*. Bloomington, IN: Solution Tree Press.

Kanold, T. D., Barnes, B., Larson, M. R., Kanold-McIntyre, J., Schuhl, S., & Toncheff, M. (2018). *Mathematics homework and grading in a PLC at Work*. Bloomington, IN: Solution Tree Press.

Kanold, T. D., Kanold-McIntyre, J., Larson, M. R., Barnes, B., Schuhl, S., & Toncheff, M. (2018). *Mathematics instruction and tasks in a PLC at Work*. Bloomington, IN: Solution Tree Press.

Kanold, T. D. (Ed.), & Larson, M. R. (2012). *Common Core mathematics in a PLC at Work, leader's guide*. Bloomington, IN: Solution Tree Press.

Kanold, T. D., Schuhl, S., Larson, M. R., Barnes, B., Kanold-McIntyre, J., & Toncheff, M. (2018). *Mathematics assessment and intervention in a PLC at Work*. Bloomington, IN: Solution Tree Press.

Kaplinsky, R. (2016, August 15). *#ObserveMe* [Blog post]. Accessed at http://robertkaplinsky.com/observeme on November 13, 2017.

Kramer, S. V., & Schuhl, S. (2017). *School improvement for all: A how-to guide for doing the right work*. Bloomington, IN: Solution Tree Press.

Lencioni, P. (2002). *The five dysfunctions of a team: A leadership fable*. San Francisco: Jossey-Bass.

Marzano, R. J. (2009, October). *Using rounds to enhance teacher interaction and self-reflection: The Marzano Observational Protocol*. Accessed at www.iobservation.com/files/Marzano-Protocol-Using_Rounds1009.pdf on September 12, 2017.

Mattos, M., DuFour, R., DuFour, R., Eaker, R., & Many, T. W. (2016). *Concise answers to frequently asked questions about Professional Learning Communities at Work*. Bloomington, IN: Solution Tree Press.

Maxwell, J. C. (2005). *The 360° leader: Developing your influence from anywhere in the organization*. Nashville, TN: Thomas Nelson.

National Council of Supervisors of Mathematics. (2008). *The PRIME leadership framework: Principles and indicators for mathematics education leaders*. Bloomington, IN: Solution Tree Press.

National Council of Teachers of Mathematics. (2014). *Principles to actions: Ensuring mathematical success for all*. Reston, VA: Author.

National Council of Teachers of Mathematics. (2018). *Catalyzing change in high school mathematics: Initiating critical conversations*. Reston, VA: Author.

Reeves, D. B. (2006). *The learning leader: How to focus school improvement for better results*. Alexandria, VA: Association for Supervision and Curriculum Development.

Reiss, K. (2007). *Leadership coaching for educators: Bringing out the best in school administrators*. Thousand Oaks, CA: Corwin Press.

Saccone, S. (2009). *Relational intelligence: How leaders can expand their influence through a new way of being smart*. San Francisco: Jossey-Bass.

Smith, M. S., Steele, M. D., & Raith, M. L. (2017). *Taking action: Implementing effective mathematics teaching practices in grades 6–8*. Reston, VA: National Council of Teachers of Mathematics.

Smith, M. S., & Stein, M. K. (1998). Selecting and creating mathematical tasks: From research to practice. *Mathematics Teaching in the Middle School, 3*(5), 344–350.

Stigler, J. W., & Hiebert, J. (1999). *The teaching gap: Best ideas from the world's teachers for improving education in the classroom*. New York: Free Press.

Townsend, R. (1970). *Up the organization: How to stop the corporation from stifling people and strangling profits*. New York: Knopf.

Williams, K. C. (2010, October 25). *Do we have team norms or "nice to knows"? Roadblock: Lack of accountability protocol* [Blog post]. Accessed at www.allthingsplc.info/blog/view/90/do-we-have-team-norms-or-nice-to-knows on June 9, 2017.

Index

professional learning communities (PLCs)
    commitment to, 10, 14, 21–24
    critical questions, 2, 4, 25
    equity and, 1–2
    pillars of, 28–31
professional learning communities framework,
    mathematics in, 3, 4
protocols. *See* assessments
psychological safety, 32
purpose (mission and vision), 10, 27, 28, 29, 30, 31

R

rating and flection framework, 109–110
Reeves, D. B., 17
reflect, refine, and act cycle, 2
    collaboration and, 68–70
    evidence of, 109–110
    leadership practices and, 24
    lesson studies and, 99
Reiss, K., 19
relational (emotional) intelligence, 10, 14, 17–18
resiliency, 21

S

Schuhl, S., 83
servant leadership, 15
singletons, collaborative teams for, 3
site-level leader
    collaboration and, 63, 67
    leadership practices and, 13
    leadership strategies and, 26

role of, 8, 9
SMART goals, 10, 27, 28, 29, 36–42, 68
Steinberg, R., 61
Stigler, J. W., 89, 93
student thinking reasoning protocol, 84, 85–86, 88
student work protocol, 77–79

T

Teaching Gap: Best Ideas From the World's Teachers for Improving Education in the Classroom (Stigler and Hiebert), 93
team artifacts, 46, 54
team-building protocol worksheet, 45
team leader
    collaboration and, 63, 67
    leadership practices and, 13
    leadership strategies and, 26
    role of, 8, 9
Thames, M. H., 89
Tier 2 interventions, 83
tight-loose leadership, 10, 42–46
Toncheff, M., 20, 23, 25, 28, 52, 67, 78
Townsend, R., 15
trust, environment of, 10, 14, 15–16

V

values (commitments), collective, 10, 27, 28, 29, 30, 31–35
vision (common purpose), 10, 27, 28, 29, 30, 31

W

Williams, K. C., 34

Every Student Can Learn Mathematics series
Timothy D. Kanold et al.
Discover a comprehensive PLC at Work® approach to achieving mathematics success in K–12 classrooms. Each book offers two teacher team or coaching actions that empower teams to reflect on and refine current practices and routines based on high-quality, research-affirmed criteria.
BKF823 BKF824 BKF825 BKF826

Mathematics Assessment and Intervention in a PLC at Work®
Timothy D. Kanold, Sarah Schuhl, Matthew R. Larson, Bill Barnes, Jessica Kanold-McIntyre, and Mona Toncheff
Harness the power of assessment to inspire mathematics learning. This user-friendly resource shows how to develop high-quality common assessments, and effectively use the assessments for formative learning and intervention. The book features unit samples for learning standards, sample unit exams, and more.
BKF823

Mathematics Homework and Grading in a PLC at Work®
Timothy D. Kanold, Bill Barnes, Matthew R. Larson, Jessica Kanold-McIntyre, Sarah Schuhl, and Mona Toncheff
Rely on this user-friendly resource to help you create common independent practice assignments and equitable grading practices that boost student achievement in mathematics. The book features teacher team tools and activities to inspire student achievement and perseverance.
BKF825

Mathematics Instruction and Tasks in a PLC at Work®
Timothy D. Kanold, Jessica Kanold-McIntyre, Matthew R. Larson, Bill Barnes, Sarah Schuhl, Mona Toncheff
Improve your students' comprehension and perseverance in mathematics. This user-friendly resource will help you identify high-quality lesson-design elements and then show you how to implement them within your classroom. The book features sample lesson templates, online resources for instructional support, and more.
BKF824

GL⦿BAL PD

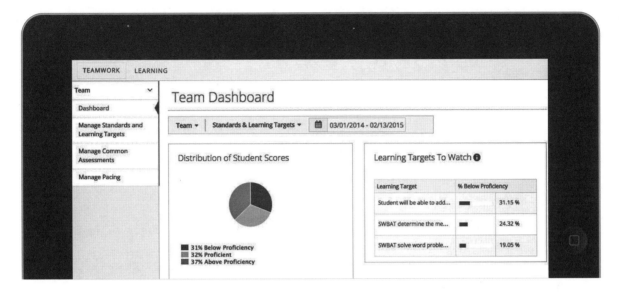

The **Power to Improve**
Is in Your Hands

Global PD gives educators focused and goals-oriented training from top experts. You can rely on this innovative online tool to improve instruction in every classroom.

- Get unlimited, on-demand access to guided video and book content from top Solution Tree authors.

- Improve practices with personalized virtual coaching from PLC-certified trainers.

- Customize learning based on skill level and time commitments.

Solution Tree

Solution Tree's mission is to advance the work of our authors. By working with the best researchers and educators worldwide, we strive to be the premier provider of innovative publishing, in-demand events, and inspired professional development designed to transform education to ensure that all students learn.

The National Council of Teachers of Mathematics is a public voice of mathematics education, supporting teachers to ensure equitable mathematics learning of the highest quality for all students through vision, leadership, professional development, and research.